CHAOS, FRACTALS, AND DYNAMICS

PURE AND APPLIED MATHEMATICS

A Program of Monographs, Textbooks, and Lecture Notes

LECTURE NOTES

IN PURE AND APPLIED MATHEMATICS

1. *N. Jacobson*, Exceptional Lie Algebras
2. *L. -Å. Lindahl and F. Poulsen*, Thin Sets in Harmonic Analysis
3. *I. Satake*, Classification Theory of Semi-Simple Algebraic Groups
4. *F. Hirzebruch, W. D. Newmann, and S. S. Koh*, Differentiable Manifolds and Quadratic Forms (out of print)
5. *I. Chavel*, Riemannian Symmetric Spaces of Rank One (out of print)
6. *R. B. Burckel*, Characterization of C(X) Among Its Subalgebras
7. *B. R. McDonald, A. R. Magid, and K. C. Smith*, Ring Theory: Proceedings of the Oklahoma Conference
8. *Y.-T. Siu*, Techniques of Extension on Analytic Objects
9. *S. R. Caradus, W. E. Pfaffenberger, and B. Yood*, Calkin Algebras and Algebras of Operators on Banach Spaces
10. *E. O. Roxin, P.-T. Liu, and R. L. Sternberg*, Differential Games and Control Theory
11. *M. Orzech and C. Small*, The Brauer Group of Commutative Rings
12. *S. Thomeier*, Topology and Its Applications
13. *J. M. Lopez and K. A. Ross*, Sidon Sets
14. *W. W. Comfort and S. Negrepontis*, Continuous Pseudometrics
15. *K. McKennon and J. M. Robertson*, Locally Convex Spaces
16. *M. Carmeli and S. Malin*, Representations of the Rotation and Lorentz Groups: An Introduction
17. *G. B. Seligman*, Rational Methods in Lie Algebras
18. *D. G. de Figueiredo*, Functional Analysis: Proceedings of the Brazilian Mathematical Society Symposium
19. *L. Cesari, R. Kannan, and J. D. Schuur*, Nonlinear Functional Analysis and Differential Equations: Proceedings of the Michigan State University Conference
20. *J. J. Schäffer*, Geometry of Spheres in Normed Spaces
21. *K. Yano and M. Kon*, Anti-Invariant Submanifolds
22. *W. V. Vasconcelos*, The Rings of Dimension Two
23. *R. E. Chandler*, Hausdorff Compactifications
24. *S. P. Franklin and B. V. S. Thomas*, Topology: Proceedings of the Memphis State University Conference
25. *S. K. Jain*, Ring Theory: Proceedings of the Ohio University Conference
26. *B. R. McDonald and R. A. Morris*, Ring Theory II: Proceedings of the Second Oklahoma Conference
27. *R. B. Mura and A. Rhemtulla*, Orderable Groups
28. *J. R. Graef*, Stability of Dynamical Systems: Theory and Applications
29. *H.-C. Wang*, Homogeneous Branch Algebras
30. *E. O. Roxin, P.-T. Liu, and R. L. Sternberg*, Differential Games and Control Theory II
31. *R. D. Porter*, Introduction to Fibre Bundles
32. *M. Altman*, Contractors and Contractor Directions Theory and Applications
33. *J. S. Golan*, Decomposition and Dimension in Module Categories
34. *G. Fairweather*, Finite Element Galerkin Methods for Differential Equations
35. *J. D. Sally*, Numbers of Generators of Ideals in Local Rings
36. *S S. Miller*, Complex Analysis: Proceedings of the S.U.N.Y. Brockport Conference
37. *R. Gordon*, Representation Theory of Algebras: Proceedings of the Philadelphia Conference
38. *M. Goto and F. D. Grosshans*, Semisimple Lie Algebras
39. *A. I. Arruda, N. C. A. da Costa, and R. Chuaqui*, Mathematical Logic: Proceedings of the First Brazilian Conference

Other Volumes in Preparation

CHAOS, FRACTALS, AND DYNAMICS

Edited by

P. FISCHER
WILLIAM R. SMITH

University of Guelph
Guelph, Ontario
Canada

MARCEL DEKKER, INC. New York and Basel

Library of Congress Cataloging in Publication Data
Main entry under title:

Chaos, fractals, and dynamics.

   (Lecture notes in pure and applied mathematics ; 98)
   Includes index.
   1. Differentiable dynamical systems--Addresses,
essays, lectures.   2. Chaotic behavior in systems--
Addresses, essays, lectures.   3. Fractals--Addresses,
essays, lectures.   I. Fischer, P. [date].   II. Smith,
William R. (William Robert) [date].   III. Series:
Lecture notes in pure and applied mathematics ; v. 98
QA614.8.C53 1985   515.3'5   85-4526
ISBN 0-8247-7325-X

MARCEL DEKKER, INC.

270 Madison Avenue, New York, New York 10016

Current printing (last digit)
10 9 8 7 6 5 4 3 2

PRINTED IN THE UNITED STATES OF AMERICA

Preface

Dynamical systems theory is the principal theme of this volume, together with related studies of chaos and fractals . Dynamical systems theory has enjoyed a recent renaissance in mathematics, dating from Lorenz's classic experimental paper in 1962. The simple nonlinear system of three differential equations he studied as a model of boundary layer convection exhibits remarkably complex behavior. In fact, similar mathematical phenomena were known to Poincaré and Birkhoff over seventy years ago, but they have received relatively little attention until recently.

"Chaos" is a term often used to describe the complicated behavior of nonlinear dynamical systems, and which is not always precisely defined in the literature. Nonetheless, this term has gained widespread acceptance and we make no apologies for its use in the title of this book. Perhaps one of the main reasons for the use of the term chaos is due to the fact that the behavior of systems exhibiting it is not yet generally well understood. It may be that, as our understanding develops, at least some systems described by that term will become more precisely classified.

"Fractals" is a term introduced by B. Mandelbrot over a decade ago to describe and classify many "irregular" shapes and patterns whose fractal (Hausdorff) dimension is not an integer. The term is also useful in describing certain aspects of dynamical systems exhibiting irregular (chaotic) behavior. Hence there has been a resurgence of interest in early studies of recursively defined systems, especially those due to Fatou and Julia in the early part of this century. These types of systems are well represented in this volume.

The behavior described by Lorenz has been found to be ubiquitous, arising in many fields, including chemistry, biology, and physics, among others. Indeed, particular phenomena from these fields can and have been used to provide prototype examples of mathematically interesting nonlinear dynamical systems. In fact, the resurgence of interest in the early part of the last decade in this type of behavior exhibited by onedimensional maps is closely associated with the biologically motivated studies of Robert May.

Another reason for the resurgence is associated with the development of digital computers, which have provided researchers with an additonal tool to study aspects of dynamical systems previously considered too complex. Indeed, many of the fractal configurations shown in parts of this book were generated by computer calculations of underlying twodimensional maps.

The papers in this volume were all presented at or resulted directly from two conferences held at the University of Guelph in March of 1981 and 1983, each entitled "Chaos Days at Guelph." The division of the book into Part I and II corresponds to the two meetings.

P. Fischer

William R. Smith

Contents

Contributors

RALPH H. ABRAHAM Mathematics Board, University of California, Santa Cruz, California

M. E. ALEXANDER Institute of Computer Science, University of Guelph, Guelph, Ontario, Canada

J. BRINDLEY School of Mathematics, University of Leeds, Leeds, England

SHUI-NEE CHOW Mathematics Department, Michigan State University, East Lansing, Michigan

P. FISCHER Department of Mathematics and Statistics, University of Guelph, Guelph, Ontario, Canada

DAVID GREEN, JR. Science and Mathematics Department, GMI Engineering & Management Institute, Flint, Michigan

OKAN GUREL IBM Cambridge Scientific Center, Cambridge, Massachusetts

MORRIS W. HIRSCH Department of Mathematics, University of California, Berkeley, California

GIKŌ IKEGAMI* Department of Electrical Engineering, University of Waterloo, Waterloo, Ontario, Canada

HÜSEYIN KOÇAK Lefschetz Center for Dynamical Systems, Brown University, Providence, Rhode Island

*Permanent address: Department of Mathematics, Nagoya University, Nagoya, Japan

W. F. LANGFORD* Department of Mathematics, McGill University, Montreal, Quebec, Canada

BENOIT B. MANDELBROT IBM Thomas J. Watson Research Center, Yorktown Heights, New York

I. M. MOROZ† School of Mathematics, University of Leeds, Leeds, England

OTTO E. RÖSSLER Institute for Physical and Theoretical Chemistry, University of Tübingen, Tübingen, Federal Republic of Germany

KATHERINE A. SCOTT Computer Center, University of California, Santa Cruz, California

CHRISTOPHER C. SHAW Division of Natural Sciences, University of California, Santa Cruz, California

WILLIAM R. SMITH Department of Mathematics and Statistics, University of Guelph, Guelph, Ontario, Canada

─────────
*Present address: Department of Mathematics and Statistics, University of Guelph, Guelph, Ontario, Canada

†Present address: School of Mathematics and Physics, University of East Anglia, Norfolk, England

CHAOS, FRACTALS, AND DYNAMICS

PART I

1

Chaostrophes, Intermittency, and Noise

Ralph H. Abraham

Mathematics Board
University of California
Santa Cruz, California

Dedicated to René Thom

In 1972, we proposed the blue sky catastrophe for periodic limit sets. Here, we describe one for chaotic limit sets. This provides a pathway to chaos quite different from the usual ones, which are all sequences of subtle bifurcations. Further, models for intermittency and noise amplification are given, based on hysteresis loops in a serially coupled chain of dynamical schemes.

PART A. SUBTLE AND CATASTROPHIC BIFURCATIONS

The classification of bifurcations into these two types was suggested in 1966 by Thom(1972), and given explicit treatment (under the names leaps and wobbles) by the author (Abraham, 1976). In this part we have two goals: to define catastrophic borders in the control space of a complex dynamical scheme, and to discuss an example of a chaostrophe, that is, a catastrophic border for the domain of a chaotic attractor, in the context of a serial chain of three oscillators.

A1. PARTITIONS AND BORDERS

We consider a vector field depending upon a parameter, also known as a metabolic field, or dynamical scheme. Let C and M be manifolds of finite dimension, $\underline{X}(\underline{M})$ a space of vectorfields on $\underline{M}$, and $\underline{F}$: $\underline{C}$ → $\underline{X}(\underline{M})$ the dynamical scheme. If $\underline{B}(\underline{M})$ is the subset of $\underline{X}(\underline{M})$ consisting of structurally unstable vectorfields, then the bifurcation set of the scheme, $\underline{B}$, is the inverse image of $\underline{B}(\underline{M})$ under $\underline{F}$.

Imagining the phase portrait of $\underline{F}(\underline{c})$ in $\{\underline{c}\}x\underline{M}$ for each $\underline{c}$ in $\underline{C}$ creates a control-phase portrait of $\underline{F}$ in $\underline{C}x\underline{M}$. We wish to concentrate on the attractors (in the sense of probability, for example) in this portrait, along with their basins and separators (the complements of the basins, elsewhere called separatrices). Let $\underline{A}$ denote the locus of attraction, the union of all the attractors of the scheme, and $\underline{S}$ denote the locus of separation, the union of all the separators of the scheme.

A relatively open subset of the locus of attraction will be called an attractrix. This is usally called a branch of the attractive surface in static catastrophe theory. A relatively open subset of the locus of separation, similarly, will be called a separatrix. This is also known as a branch of the repelling surface in static catastrophe theory.

We assume that the scheme, $\underline{F}$, is generic in any reasonable sense. Specifically, it is as transversal to $\underline{B}(\underline{M})$ as possible, and over each point $\underline{b}$ in the bifurcation set, there is a single bifurcation event in the phase portrait of $\underline{F}(\underline{b})$. We see in examples that this bifurcation event normally involves a single attractrix and a single separatrix, or it involves no attractrix. Thus, $\underline{B}$ may be divided in two parts. Here, we will be interested in the attractrix bifurcations only, in which an attractrix and a separatrix are involved. Further, we will discuss only the hypersurfaces contained in this part of the bifurcation set, which we call the attractrix bifurcation hypersurfaces in the control manifold, $\underline{C}$. And finally, these may be isolated hypersurfaces in the bifurcation set, or they may be hypersurfaces of accumulation, from one or both sides. We will refer to one of these isolated attractrix bifurcation hypersurfaces as a border, if an attractor appears or disappears during the bifurcation occuring across it. Otherwise, we call it a partition. The borders belong to the boundaries of the domains of attraction, the regions of control space in which certain attractors exist. These domains are the shadows (images in $\underline{C}$ under the projection from $\underline{C}$ x $\underline{M}$ onto the first factor) of attractrices, and the borders are shadows of boundaries of attractrices. Borders may always be oriented, by a normal vectorfield pointing toward the exterior of the region it bounds. Partitions belong to the interiors of the domains of attraction, and may radiate inward from a border. Precise definitions are given in Section A5.

A2. STANDARD EXAMPLES WITH ONE CONTROL

Specializing the preceding definitions to the case in which the control space, $\underline{C}$, is a line or circle, yields the most important examples. Thus, $\underline{F}$: $\underline{C} \rightarrow \underline{X}(\underline{M})$ is a generic arc or generic loop, the bifurcation set, $\underline{B}$, is zero-dimensional, every point is a hypersurface, and we fasten attention upon the isolated points at which an attractrix appears or disappears. These are the borders in this context.

In case the dimension of the state space, $\underline{M}$, is two, everything is known about the bifurcations of generic arcs. The attractrices correspond to static or periodic attractors. The separatrices are generated by the insets of limit points and cycles of saddle type. The isolated points of the attractrix bifurcation set belong to a known list of possible models, while the accumulation points (also called thick bifurcations, Abraham and Shaw, 1983) are due to a single phenomena: non-trivial recurrence on a torus.

The different types of known bifurcation obviously fall into two categories, subtle and catastrophic. The catastrophic ones are the borders, while the subtle ones are the partitions. This classification is given in Table 1. Drawings of the locus of attraction, in most cases, may be found elsewhere (Abraham and Marsden, 1978, Abraham and Shaw, 1983). An exception is solidification, a type of Hopf bifurcation, which is shown in Figure 1.

Another of these, the periodic blue sky event, will be described in detail in the next section.

If the dimension of the state space, $\underline{M}$, is three or more, then chaotic attractorts may (and usually do) occur. The full list of these objects is not yet known, even in three dimensions. Their bifurcations, which include pathways to chaos, are just beginning to be discovered by

TABLE 1. ATTRACTRIX BIFURCATIONS OF GENERIC ARCS

CATASTROPHIC BORDERS	SUBTLE PARTITIONS
Static creation	
Static solidification	Hopf excitation
Periodic creation	
Periodic solidification	Neimark excitation
Murder	Subharmonic division
Periodic blue sky catastrophe	

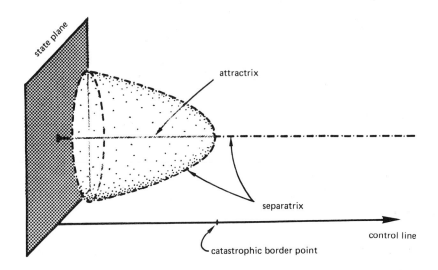

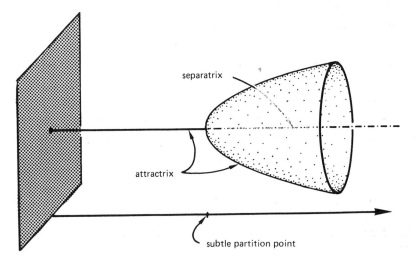

FIGURE 1.   SUBTLE AND CATASTROPHIC HOPF BIFURCATIONS.
The dashed curves repel, solid attracts.

experimental dynamicists. Yet we presume that these also will fall into the two categories, subtle and catastrophic. So far, the examples known are primarily of the subtle sort. (For some exceptions, see Arneodo, Coullet and Tresser, 1980; and Grebogi, Ott, and Yorke, 1982.)

A3. THE BLUE CYCLE PERIOSTROPHE.

Recall that in the dynamic annihilation catastrophe, a periodic attractor (attractive limit cycle, oscillation) vanishes. It collides with a limit cycle contained in its separator. Its attractrix meets its separatrix. In 1972, we conjectured the existence of a blue sky catastrophe, in the context of a dynamical scheme (Abraham, 1972). As in dynamic annihilation, a limit cycle would disappear into the blue sky. But in this case, it would not be cancelled through collision with another limit cycle. Instead, its period (length of its time cycle) would become infinite. It would just slow down, and cease to oscillate.

In the course of time, this conjecture was confirmed (Takens, 1974, Devaney, 1977). In the blue sky event now well known, expressed in the case in which the disappearing limit cycle is an attractor, a periodic attractor just slows down and stops. But in fact, at the moment of disappearing into the blue, it does indeed collide with another trajectory, also an oscillation of infinite period. This is a homoclinic trajectory, or saddle self-connection, associated with its separatrix. This also provides an illustration of basin catastrophe, as the basin of the periodic attractor vanishes at the moment of periostrophe, along with its (possibly unbounded) tail. The event is shown in Figure 2.

A4. THE BLUE BAGEL CHAOSTROPHE

This event will be constructed from the blue cycle periostrophe by Cartesian product with a circle plus a perturbation, to obtain a generic arc with known behavior. In other words, we perturb a blue cycle scheme with a forcing oscillation. Supposing the state space of the original scheme to be a plane, as shown in Figure 3, the forced scheme will have a solid ring for its state space, as shown in Figure 3. The plane within this ring corresponding to phase zero of the driving oscillation will be useful in our discussion. We call it the strobe plane (Abraham and Shaw, 1982).

Before the bifurcation of the forced scheme, near the border point of the original blue cycle scheme, we have an attractive torus. The torus

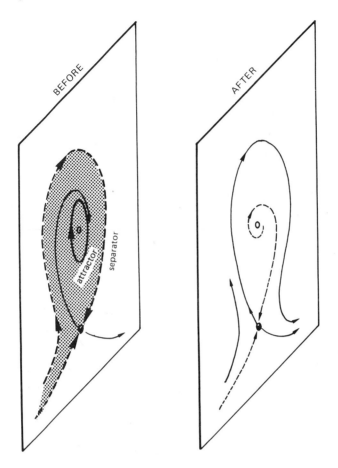

FIGURE 2. THE BLUE CYCLE PERIOSTROPHE.

meets the strobe plane in the periodic attractor of the original blue
cycle scheme, the one which vanishes into the blue. This torus contains a
braid of periodic attractors. Their basins, within the invariant torus,
are separated by a complementary braid of periodic trajectories which are
repelling, within the torus. The saddle point of the original scheme
becomes a limit cycle of saddle type in the ring model (state space) of
the combined scheme. The inset of this limit cycle is a scrolled
cylinder, generated by the inset curve of the original scheme, visualized
on the strobe plane. Likewise, the outset of the limit cycle is another,
complementary, scrolled cylinder.

After the bifurcation event is over, well beyond the border point of
the original scheme, the attractive torus is gone, braids and all,

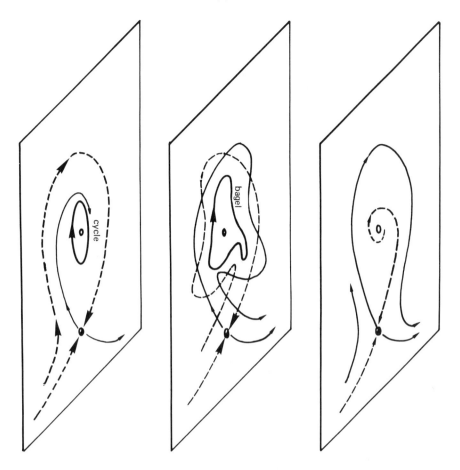

FIGURE 3. THE BLUE BAGEL CHAOSTROPHE, IN STROBE PLANE SECTIONS.

vanished into the blue yonder. The periodic trajectory of saddle type, along with its complementary scrolled cylinders, remains. But the relationship between them is reversed.

During the bifurcation event, they have crossed. The crossing does not occur at a single bifurcation value of the control variable, for we have perturbed the forced oscillation into a generic arc. Thus the inset and outset cylinders, visualized as curves within the strobe plane, must pass through each other within an interval of homoclinic transversal intersection. We call this the homoclinic interval. There are many possiblities for the prolonged passage, because of the likelihood of Birkhoff rechambering bifucations (Abraham and Marsden, 1978). One of the

simplest is indicated in Figure 3. <u>At some point in the homoclinic</u>
<u>interval, the attractive torus disappears, braids and all.</u>

So far, we have described aspects of this generic arc which are
mathematically known. But now, we venture into <u>conjecture</u>:

Within the homoclinic interval, the <u>tangled inset</u> of the periodic
saddle behaves as a repellor, and as the separator for the basin of the
blue-bound attractive torus. Eventually, the torus will collide with this
tangled inset. But before this happens, the convolutions of the tangled
cylinder mold the torus into a form much like the chaotic attractor found
by Rob Shaw. Discovered in experiments with the forced Van der Pol
scheme, this object looks like a very dog-eared bagel (Abraham and Shaw,
1982). Thus, at some point within the homoclinic orbit, there is a subtle
bifurcation, where the attractive torus becomes an attractive, chaotic
bagel. <u>At the final endpoint of the homoclinic interval, the chaotic</u>
<u>bagel collides with the tangled inset, now tangent to the outset, of the</u>
<u>periodic saddle, and vanishes into the blue.</u>

This is an example of a <u>chaostrophe</u>, as the chaotic bagel attractor
has disappeared discontinuously. It is also an example, in the context of
three-dimensional dynamical systems, of the <u>fractal torus crisis</u> described
for three dimensional maps by Grebogi, Ott, and Yorke (1982).

We may end this fantasy with the further conjecture, that this event
may be found experimentally in the forced Van der Pol scheme.

A5. CATASTROPHES WITH SEVERAL CONTROLS

Now, we approach our second goal in this part, the definition of <u>subtle</u>
<u>partitions</u> and <u>catastrophic borders</u>, in the general context of a dynamical
scheme with several control paramters. In review, the control space, $\underline{C}$,
is a finite-dimensional manifold. The dynamical scheme, $\underline{F}: \underline{C} \to \underline{X}(\underline{M})$ is
assumed to be <u>generic</u> in a sense we have not made precise. The
bifurcation set, $\underline{B}$, is a subset of $\underline{C}$. We consider a subset $\underline{H}$ of $\underline{B}$ which
is an oriented hypersurface of $\underline{C}$, and which is <u>isolated</u> in the sense that
a neighborhood of $\underline{H}$ in $\underline{C}$ intersects $\underline{B}$ only in $\underline{H}$. (In what follows, it may
only be essential that the hypersurfaces be isolated on one side.) Under
these assumptions, the isolated hypersurface corresponds to a single
bifurcation event in the portrait of the dynamical scheme.

Finally, the hypersurface is a <u>border</u> if this event involves either
the appearance or the disappearance of an attractor, and the hypersurface
is oriented toward the exterior of the domain of this attractrix.
Otherwise, the hypersurface is a <u>partition</u>.

We now define a generic sub-arc of a dynamical scheme, which will facilitate a more precise distinction between subtle partitions and catastrophic border in this context.

As a hypersurface of $\underline{C}$, $\underline{H}$ is of codimension one. Any curve $\underline{d}$: $\underline{I} \rightarrow \underline{C}$ in $\underline{C}$ (where $\underline{I}$ is an open interval of real numbers) which is transversal to $\underline{H}$ may be composed with the scheme $\underline{F}$ to obtain an arc, $\underline{f} = \underline{F} \circ \underline{d}$: $\underline{I} \rightarrow \underline{X}(\underline{M})$. This is a one-parameter dynamical scheme, the sort discussed in Section A2. As $\underline{F}$ is generic, so is $\underline{f}$. We may refer to such a generic arc, derived from the dynamical scheme $\underline{F}$, by composition with a curve transversal to its bifurcation set, as a generic sub-arc.

Note that the bifurcation set of a generic sub-arc cutting $\underline{H}$ transversally at the point $\underline{h}$ (at least, if it is sufficiently short) consists of the single point, $\underline{h}$. Finally, we assume the short, generic sub-arc crosses the border in the direction of the outward normal, or orientation. We call this a transverse at $\underline{h}$.

This is the auxiliary notion we need for the definition of subtle partitions and catastrophic borders in this multi-dimensional context.

Here at last is the definition. Every point $\underline{h}$ in the hypersurface, $\underline{H}$, is either subtle or catastrophic. It is subtle if, roughly, the locus of attraction is continuous over it. More precisely, the point $\underline{h}$ is a subtle partition point if every sub-arc transverse at $\underline{h}$ has a subtle bifurcation at $\underline{h}$, in the sense of Section A2. Otherwise, the point $\underline{h}$ is a catastrophic border point. We use the word catastrophe at once for the point, $\underline{h}$, in the bifurcation set within the control space, and the discontinuity in the affected attractrix.

A subset of an isolated, oriented, attractrix bifurcation hypersurface consisting entirely of catastrophe points is called a catastrophic border, or just plain border. We define subtle partition, or partition, similarly. Every such hypersurface may be decomposed into a union of catastrophic borders and subtle partitions.

So much for the definition. A better idea of the distinction between a subtle bifurcation and a catastrophe may be gleaned from the example shown in Figure 4. Derived from the Andronov–Takens (2,–) model by symmetry-breaking, it is four-dimensional (Takens, 1974; Abraham and Marsden, 1978). Both $\underline{C}$ and $\underline{M}$ are planes. So we show the tableau of sample phase portraits within each region of the control plane. The borders, in this planar control space, are the four solid curves radiating from the central point. Omitting a neighborhood of this point, they are all isolated, and involve an attratrix catastrophe. Two of these are

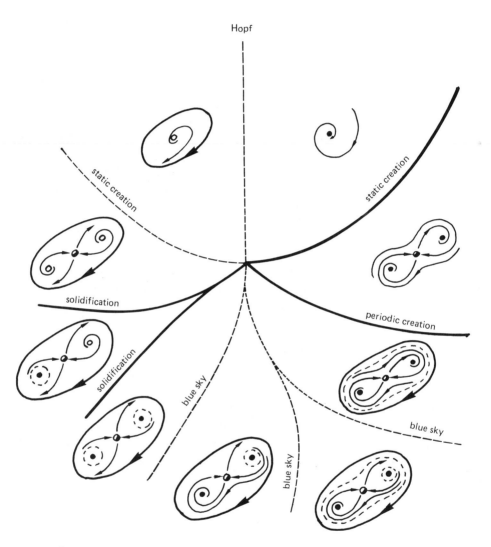

FIGURE 4. SUBTLE PARTITIONS AND CATASTROPHIC BORDERS.
Subtle partitions are represented by broken curves,
catastrophic borders by solid curves.

solidifications. The other two are a static creation, and a periodic
creation. The remaining five curves radiating from the central
bifurcation point are subtle partitions.

PART B. HYSTERESIS, HOLONOMY AND NOISE.

Thom has written that the primary events in morphogenesis are the
catastrophic bifurcations, and we share this view. In this part, we
indicate some subtleties of catastrophic borders in the context of
serially coupled chains of dynamical schemes. In particular, a chain of
three oscillators serves as our standard example.

B1. HYSTERESIS WITH ONE CONTROL

Hysteresis refers to the failure of a system to return to its original
state, after a temporary change of its controls. We may interpret this in
the context of complex dynamical system theory (Abraham, 1983a) as
follows.

We consider three simple dynamical schemes, the output of one
determining the control parameters of another, in a chain. This is an
example of a serial chain, and is shown in Fig. 5(a), with the standard
convention of complex dynamics: the solid dots represent the component
schemes, while the hollow dots denote the serial coupling functions. We
will call the first scheme the master controller, and the whole serial
chain driven by it the slave chain. This is exemplified by the classical
model of Lord Rayleigh for forced oscillation, in which both systems of
the slave chain are running in periodic attractors. We assume now that
the master controller is an oscillator, relatively slow with respect to
the dynamics slave chain.

This forced oscillation is an hysterical, (or, each cycle is an
hysteresis loop) if the slave scheme (as a coupled dynamical system) does
not return to the same attractor after each period of the forcing
oscillation. Hysteresis is characteristic of serial chains.

We now illustrate this phenomenon in a system with a single master
control. An early example, found by Duffing in Lord Rayleigh's model for
forced oscillations of the damped harmonic oscillator, is shown in Figure
5 (Abraham and Shaw, 1982). This is a periodic version of a configuration
common in elementary catastrophe theory, which may be called the periodic
double fold.

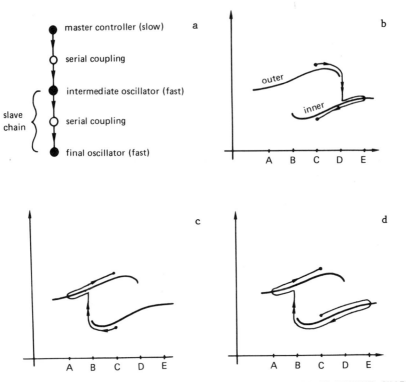

FIGURE 5. HYSTERESIS LOOPS OF DUFFING IN A THREE COMPONENT CHAIN.

This is the portrait of the slave chain, a forced oscillator. Here there is one control, corresponding to the frequency of the intermediate forcing oscillation. This control is to be determined by the master oscillator. The attractors are periodic, the separator is periodic, the bifurcation set in control space consists of two points (B and D). Both borders, the two <u>attractrices</u> (solid surfaces) overlap in the interval between these points, and in this interval they are separated by the separatrix (shaded surface).

Here are some exemplary hysteresis loops. (1) If the coupled (slave) oscillator is on the outer attractrix (oscillation) over control C, the cycle CEC in control space will leave it on the inner surface, as shown in Fig. 5(b). (2) Starting again on the inner attractrix over point C, the cycle CAC will return the slave system to the outer oscillation, as shown in Fig. 5(c). (3) The compound control cycle CECAC will not change the outer oscillation, but will change the inner oscillation to the outer, as shown in Fig. 5(d). We consider this compound cycle to be a hysterersis loop also.

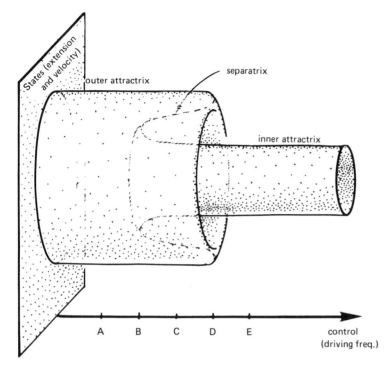

FIGURE 5 (cont.)

The maps from the set of attractors over C into itself, determined by all hysteresis loops at C, comprise what we have called (Abraham, 1983b) the <u>holonomy monoid</u> of C. It is clear in this example that the parameterizations of the hysteresis loops do not affect the holonomy maps. But in most cases, parameterization does affect the holonomy, as we shall see in Section B3.

B2. HYSTERESIS WITH TWO CONTROLS

The simplest example of hysteresis with two controls (two-dimensional control space of the driven system) is provided by the static cusp catastrophe of elementary catastrophe theory. This is essentially three-dimensional, and thus is easily visualized. A periodic version, the <u>periodic cusp catastrophe</u>, is essentially four-dimensional. One three-dimensional section is identical to Figure 5. But viewing the strobe-zero plane of one of the periodic attractors in place of the planar state space, we obtain another three-dimensional section of the four-dimensional diagram, as shown in Figure 6. Here, the borders are the two curves of

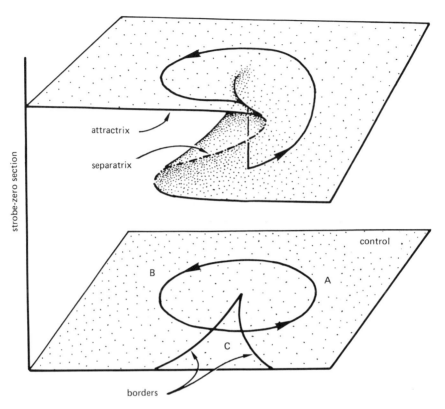

strobe-zero section

attractrix

separatrix

control

B

A

C

borders

FIGURE 6. PERIODIC CUSP CATASTROPHE.

the cusp, in the control plane. The attractrices, actually three-dimensional hypersurfaces in four-dimensional space, appear in the section as surfaces.

We consider a closed curve in the control space, as shown, oriented CABC. This may be the image of a periodic attractor of the master system, a forcing oscillator. As this path crossed borders, it is a hysteresis loop. Its holonomy map takes the lower oscillation attractrix to the upper one, which is left fixed by the map. As in the preceding section, the parameterization of this path does not affect its holonomy.

This diagram occurs in Lord Rayleigh's model for the forced oscillator, as Duffing discovered, if the amplitude of the intermediate oscillator is controlled by the master, as well as its frequency. Typical output, represented as a time series of a single state variable of the final, driven oscillator, is shown in Figure 7. An interesting application to memory has been made by Zeeman (1977).

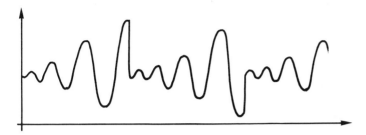

FIGURE 7. TIME SERIES OF AN HYSTERICAL CYCLE.

B3. STOCHASTIC HOLONOMY

Here is another example of hysteresis with two controls. It is extracted
from Figure 4. This is the portrait of another forced oscillator scheme,
which will again play the role of the slave system. Adding a closed curve
to the control space, enclosing the central point of bifurcation, we
consider the holonomy of this curve. A master oscillator may be imagined,
driving the controls of the slave system around this closed curve. This
time, the parameterization of the curve will affect the holonomy map. For
ease of visualization, we replace the control space by this curve, so that
the restricted diagram becomes three-dimensional. That is, we visualize
the Cartesian product of the planar state space and the control cycle.
Ignoring subtle bifurcations, the result is shown in Figure 8. (See also
Abraham and Shaw, 1983.) Beginning at point B on the control cycle, there
is a single attractor, a point. Moving counterclockwise, this soon
becomes periodic, by a Hopf bifurcation. But this is subtle, we pay no
attention. Later, there is a static creation event, but our attractor is
not affected. Two repellors and a saddle are created, all static.

At last, there is a border at point x. One of the point repellors
has a Hopf bifurcation, creates a nearby periodic repellor, and itself
becomes a point attractor. This is a solidification. So at point C,
there are two attractors in competition, one static and one periodic.
Soon, at point y, the other point repellor solidifies also. Now there are
three attractors in competition, at point D.

There follow three blue sky bifurcations. But these involve the
appearance out of the blue of periodic repellors. So, although they are
catastrophic, no attractor is affected. We ignore these also.

At point z, there is a periodic annihilation, and the periodic
attractor disappears. This is the crucial event in this example. At

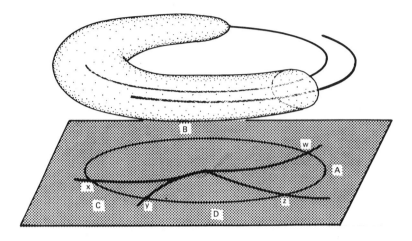

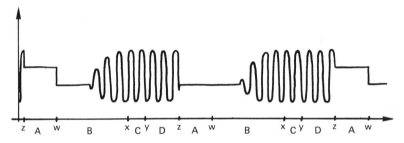

FIGURE 8. A STOCHASTIC HOLONOMY.

point A, there are only two attractors in competition, both static. The
basins are intertwined teardrops, like _yin_ _and_ _yang_.

At point w on the control cycle, there is a static annihilation
catastrophe. One of the static attractors collides with its separator and
is cancelled. One point attractor is left, and we are back at the
beginning of the cycle, at point B.

Finally, we compute the holonomy of this curve, but beginning at
point A. We may start with either of the static attractors existing at
this point, which we shall call _yin_ and _yang_. At the end of one period of
the driving cycle, the driven system will be in one or the other of these
states. The same result will be obtained, no matter which attractor we
start with, because at point B there is only one attractive state anyway.
Thus the result of one control cycle, yin or yang, defined the holonomy of
this closed curve, ABCDA. Which will it be?

All is simple until we reach point z. Just before this time, the
system must be in the periodic attractor, even though two new attractors,

the points yin and yang, have been born. At the moment border z is crossed, the periodic attractor disappears. At this moment, the driven system is in a well-defined phase of its oscillation, and thus in some computable point in its planar state space. This point is either in the basin of yin, or it is in the basin of yang. We discount the possibility that it is in the separator at that moment, as this has probability zero. If it is in the yin basin at border crossing time, it will stay in the yin basin until the next border crossing. Thus back at point A, after a full cycle of the driving oscillation, it is still yin, and the holonomy of the closed curve is yin: both yang and yin are mapped to yin.

We see that the holonomy is computable in this example, and that it depends critically on the phase of the driven oscillation at the z-border time. Thus, a slight change in the parameterization (not necessarily changing the period of the forcing oscillation) can switch the holonomy of the curve from yin to yang!

But what would be the holonomy in case the attractor vanishing into the blue at phase z was chaotic, instead of periodic? In this case, which must occur very frequently in applications, we may only hope to calculate the probability of the yin and yang results of a holonomy, based upon the measure of the intersection of the disappearing chaotic attractor with the basins of the non-disappearing attractors. These probabilities will obviously be independent of the parameterization of the closed curve. They depend only on the borders crossed. This is what we mean by stochastic holonomy. It applies equally to the preceding example, in which the attractor vanishing at z is periodic. The stochastic holonomy, averaged over the attractor, apparently depends continuously on the parameterization of the curve. This is more useful than the exact holonomy, which depends hypersensitively on the curve.

B4. MODELS FOR INTERMITTENCY AND NOISE

We have already seen, in Figure 7(b), a time series exhibiting a typical example of intermittency. This is characteristic of serial chains, as we have indicated. In this case, the model behind the time series is made of three oscillators, serially coupled in a chain. The slave chain in this case, a forced oscillator system, is characterized by the periodic cusp catastrophe of Duffing. A related phenomenon, shown in Figure 7(a), differs in that the master cycle has been deformed, through a nontransversal intersection with the borders, so as to significantly change the holonomy of the hysterical cycle. Through training, we may learn to recognize holonomy from the observed behavior of a system, and thus to create a suitable complex dynamical model for it. We will discuss a few variations on this scheme, to give four examples of this modeling strategy.

A. In the situation illustrated in Figure 7(a), let the periodic attractor of the master system be replaced by a chaotic attractor. Then the transitions between the two periodic states of the slave chain (forced oscillator) will become aperiodic in time. The time series will appear noisy, and its spectral analysis will reveal the spectrum of the master system, with the discrete spectra of the two periodic states of the slave system superimposed. The distribution of power between these two discrete spectral sequences will be indicative of the stochastic holonomy of the hysterical master macron: the master attractor, imaged in the slave control space by the serial coupling map. This is a strategy for modeling systems displaying intermittency.

B. Suppose that without changing its stochastic holonomy type, the macron is shrunk to a very small object near the cusp. At the same time, we will imagine the two attractrices of the slave portrait to be periodic motions of the slave system of very disparate amplitudes. In this situation, the noise of the master system is amplified greatly by the slave system. This is a strategy for modeling nonlinear noise amplifiers, which contribute their own periodic sequences to the output power spectrum, but do not otherwise change the noise charateristics of the input signal.

C. Suppose the slave system is made chaotic, so that while the cusp portrait still applies, each attractrix is the locus of a chaotic bagel. The borders, comprising the curves of the cusp in the control plane, are blue bagel chaostrophes, as described in Section B4. Then periodic input from the master system is amplified to periodically intermittent noise. A long time series might permit the recognition of the bagels from their characteristic power spectra, if the master oscillation is known. However, if the master system also becomes chaotic (through the action of a fourth system on its controls, a four-component serial chain), for example by a thick bifurcation resulting in a Rössler band, then the analysis of the component systems from the output time series could be hopeless. Still, exploration of serial chain behavior through fast simulations could provide enough experience to enable a strategy to evolve for modeling complex systems such as physiological or ecological networks.

D. Replace the portrait of Figure 7 with that of Figure 8. With a periodic master system, the macron (master cycle imaged in control space of the slave system by the serial coupling map) is hysterical, with a time series such as that shown in Figure 9. Here the static attractors of the slave system, yin and yang, occur periodically. But which one occurs depends on the exact state of the trajectory of the entire serial chain at the moment the master cycle crosses the critical border (blue cycle periostrophe) of the slave system. So if we now allow the master system to become chaotic through a subtle bifurcation (again, with a four-component chain), these occurrences of yin and yang will become stochastic. Still, their average frequencies of occurrence will reveal the stochastic holonomy of the master macron in the control plane of frequency and amplitude of the slave system. In this way, some information about the component systems of a serial chain may be gleaned from a time series output from the final system of the chain.

Finally, we note that the longer the serial chain, the more difficult the analysis of its output in terms of qualitative behavior of its component systems.

ACKNOWLEDGEMENTS

These ideas have been instigated by René Thom. In 1972, inspired by his book, I visited him at Bures-sur-Yvette. He showed me a book, Kymatiks, which led to my meeting with the author, Hans Jenny, to my eventual construction of the 4 inch Jenny Macroscope in Santa Cruz in 1974, and to the many experiments with three-components serial chains (using forced oscillation in fluids) which continue even now. These experiments are the source of the ideas presented here.

It is a pleasure to thank René Thom and Hans Jenny for their inspiration in the past, Paul Kramerson for his help throughout the years, in the construction of the macroscope and in the experiments, Fred Abraham and Timothy Poston for their comments on an early draft of this paper, Christopher Shaw for drawing the figures, and the University of California and the National Science Foundation for support during this past decade.

BIBLIOGRAPHY

Ralph H. Abraham, 1972. Hamiltonian Catastrophes, Univ. Claude-Bernard, Lyons.

Ralph H. Abraham, 1976. Vibrations and the realization of form, in: E. Jantsch and C.H. Waddington, eds., Evolution and Consciousness, Addison-Wesley, Reading, MA (1976).

Abraham, R. H., 1983a. Categories of dynamical models, in: T.M. Rassias (ed.), Global Analysis-Analysis on Manifolds, Teubner, Leibzig (in press).

Abraham, R. H., 1983b. Dynamical models for thought, (preprint).

Abraham, R. H., and J. E. Marsden, 1978. Foundations of Mechanics, 2nd ed., Benjamin-Cummings, Reading, MA 08126.

Abraham, R. H. and C. D. Shaw, 1982. Dynamics, the Geometry of Behavior, Part 1: Periodic Behavior, Aerial Press, Box 1360, Santa Cruz, CA 95061.

Abraham, R. H. and C. D. Shaw, 1983. Dynamics, a visual introduction, in: F.E. Yates (ed.), Self-Organizing Systems, Plenum, (to appear).

Devaney, R. L., 1977. Blue sky catastrophes in reversible and Hamiltonian systems, Indiana Univ. Math. J. 26: 247-263.

Grebogi, C., E. Ott, and J. A. Yorke, 1982. Crises, sudden changes in choatic attractors, and transient chaos, (preprint).

Takens, F., 1974. Forced oscillations and bifurcations, Math. Inst. Comm. 3, Univ. Utrecht.

Thom. R., 1975. Structural Stability and Morphogenesis, an Outline of a General Theory of Models, Engl. tr. by D. Fowler, Benjamin-Cummings, Reading, MA, (1975).

Tresser, C., A. Arneodo, and P. Coullet, 1980. On the existence of hysteria in a transition to chaos after a single bifurcation, J. Phys. Lett. 41: L-243-L-246.

Zeeman, E. C., 1977. Catastrophe Theory, Addison-Wesley, Reading, MA: pp. 293-300.

2

The Outstructure of the Lorenz Attractor

Ralph H. Abraham

Mathematics Board
University of California
Santa Cruz, California

Christopher C. Shaw

Division of Natural Sciences
University of California
Santa Cruz, California

Dedicated to René Thom.

In the chaotic attractors we have come to know experimentally, there frequently are distinguished critical points or closed orbits which organize the geometry. Here, we describe the geometry of the Lorenz attractor in terms of a yoke of outsets from three such distinguished organizers, and speculate on the generalization of this outstructure to other chaotic attractors.

1. Neat heteroclines. Consider two basic sets of a flow, Alpha and Omega. That is, each set is hyperbolic, invariant, and indecomposable, or Axiom A. These are heteroclinic if there is a trajectory from one to the other. Let us suppose there is a trajectory from Alpha to Omega. Thus, some trajectory has Alpha for its alpha limit set and Omega for its omega limit set. Then it follows that the outset (unstable manifold) of Alpha, Out(A), approaches arbitrarily close to the outset of Omega, Out(Y). Considering the implications of the hyperbolicity of Omega, and the invariant manifold theorem, there must be an intersection of the boundary of Out(A) with Out(Y) itself. We say that Alpha is neatly heteroclinic to Omega if the dimension of Out(A) is one more than the dimension of Out(Y),

and <u>the boundary of Out(A) is identical to Out(Y).</u> This is the case in the Lorenz attractor, as we shall see.

2. <u>Yokes and Coboundaries.</u> We now suppose, for simplicity, that the basic sets under discussion are hyperbolic critical elements, that is, critical points or closed orbits. Further, we consider three of these, A, B, and Y, where both A and B are heteroclinic to Y. We call this a <u>heteroclinic yoke.</u> We will see that these yokes can behave very much like homoclinic cycles in some flows: in the presence of reinsertion, they may make horseshoes, knots, and chaos. Now suppose the yoke is neat, that is, both of the heteroclinic links are neat. Then Out(A) and Out(B) are both bounded by Out(Y). Due to the hyperbolic structure of the three critical elements, the closure of the union of the three outsets is locally attractive. It is a candidate for an attractor, in fact.

3. <u>Reinsertion.</u> Note that the three outsets of a yoke must go somewhere. The omega limit sets in the boundary of these outsets are also yoked. But in the case of a neat yoke, if we suppose that the entire boundary of Out(A) and Out(B) is Out(Y), then Out(Y) has nowhere to go. So, this is only possible if Out(Y) either goes off to infinity, or it is <u>reinserted</u>, as Rössler would say. That is, the boundary of Out(Y) is found in the closure of the union of the three outsets, the candidate attractor. And both of these cases occur in the Lorenz attractor, as we show visually in the next section.

4. <u>Example: the Lorenz attractor.</u> Here is a neat yoke, expressed in a sequence of eight drawings which we made while trying to understand Perello (1980).

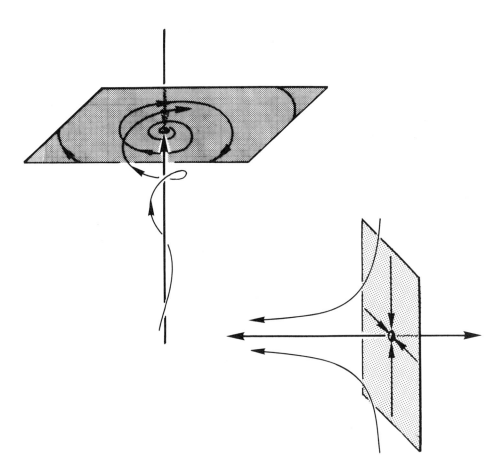

FIGURE 1. Here are two saddle points, A and Y. They are hyperbolic, in three dimensions. One, A, has index 2, with spiral dynamics on its planar outset (shaded). The other, Y, has index 1, with nodal dynamics on its planar inset (dotted), In(Y). The two outsets are attractive, as shown by the neighboring trajectories. As Out(A) and In(Y) are both two-dimensional, they could intersect transversely in three space. If they did, the transversal intersection would have to be a trajectory, the heteroclinic trajectory.

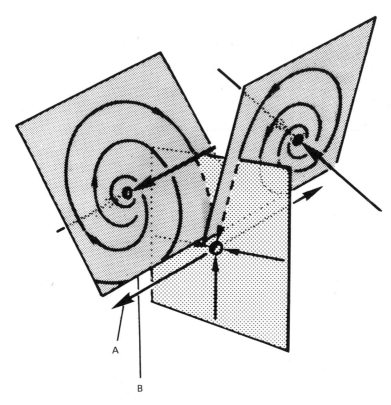

A

B

FIGURE 2. Adding another saddle point, B, essentially identical to A, we make a yoke like this. Both A and B are heteroclinic to Y. They are transversally heteroclinic, as the two planar outsets (shaded) intersect the planar inset (dotted) transversally. There are two heteroclinic trajectories in this yoke. Note that the arriving outsets are incident upon the departing outset, at Y. Thus, it is possible that this is a neat yoke. Next, we will see where these outsets end up.

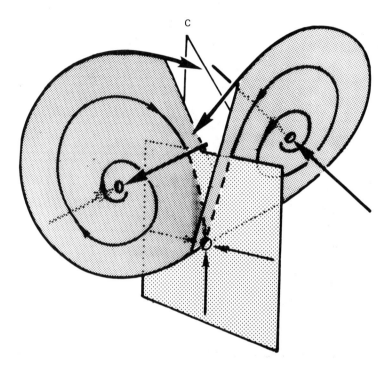

FIGURE 3. As the arriving outsets, Out(A) and Out(B), both have spiral
dynamics, the departing outset which bounds them, Out(Y), swirls around
and reinserts, as shown here. It can not go off to infinity.

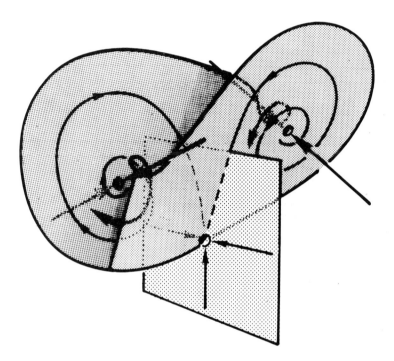

FIGURE 4. The result of reinserting is this: as each branch of Out(Y)
swirls around one of the shaded outsets, it approaches near the other
shaded outset. It gets attracted, as outsets are attractive. Thus, the
omega limit set of Out(Y) is within the closure of the union of the three
yoked outsets.

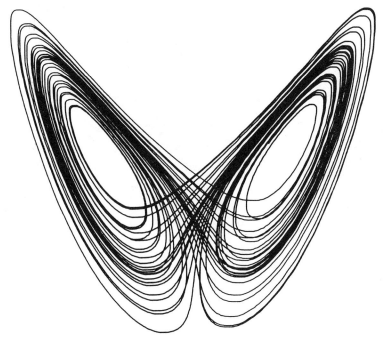

FIGURE 5. And here, for comparison, is a computer drawing of the Lorenz attractor. Inspection of the equations reveals the three distinguished saddle points, right where we want them. But the planar inset of the saddle point in the lower center is qualitatively invisible. It is a kind of separatrix. Now we will add it to the picture, with its full extension.

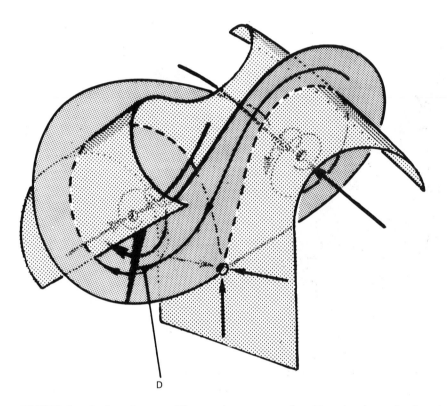

D

FIGURE 6. Referring to Figure 4, we run the flow backwards in time, to extend the planar (dotted) inset outwards from Y. It follows the heteroclinic trajectories (dashed) back to the yoked saddles, A and B, scrolling as it goes.

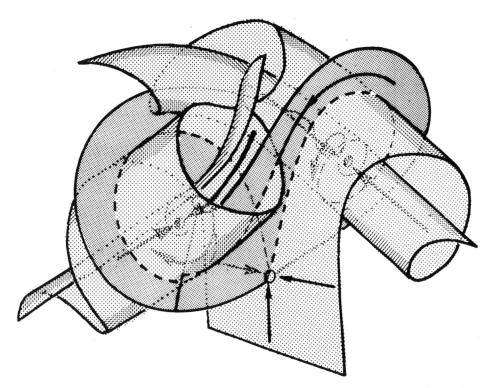

FIGURE 7. Extending the dotted inset farther backwards in time, it
scrolls up tightly around the <u>one-dimensional</u> <u>insets</u> of A and B, In(A) and
In(B). We have not said much about these curves so far. But if we could
reverse the arrow of time for a moment, we would have a neat heteroclinic
from Y to A (likewise, from Y to B) and thus In(A) and In(B) comprise the
boundary of Out(Y). We may call this a <u>neat</u> <u>reverse</u> <u>yoke</u>. The boundary
of Out(Y) also contains the repellor at infinity.

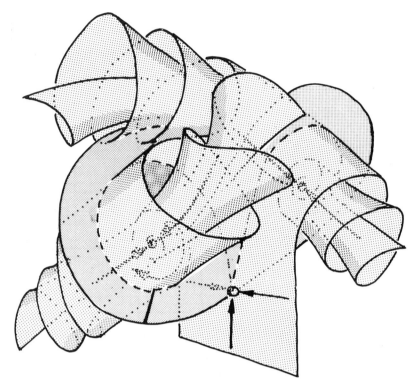

FIGURE 8. Extending the dotted inset farther backwards still, the four
ends of the scrolls are pulled out along the curves, In(A) and In(B),
toward their source at infinity.

ACKNOWLEDGEMENTS

It is a pleasure to thank Rob Shaw for contributing Figure 5, and Aerial
Press for supporting us during the preparation of this paper.

BIBLIOGRAPHY

Abraham, R. H. and C. D. Shaw, 1981. Dynamics, the Geometry of Behaviour,
Part 2: Chaotic and Stable Behavior, Aerial Press, Santa Cruz, CA 96061.

Williams, R. F. 1976. The structure of Lorenz attractors, preprint.

Gurel, O., 1981. Exploding dynamical systems, preprint.

Perello, C., 1980. Intertwining invariant manifolds and the Lorenz
attractor, in Global Theory of Dynamical Systems (Z. Nitecki and C.
Robinson, eds.), pp. 375-378, Springer, New York.

3

Chaos and Intermittency in an Endocrine System Model

Ralph H. Abraham
Mathematics Board
University of California
Santa Cruz, California

Hüseyin Koçak
Lefschetz Center for Dynamical Systems
Brown University
Providence, Rhode Island

William R. Smith
Department of Mathematics and Statistics
University of Guelph
Guelph, Ontario, Canada

A modification to a response-inhibition model for the hypothalamic-pituitary-gonadal axis of the male reproductive system gives rise to two periodic attractors in a bifurcation diagram exhibiting hysteresis and intermittency. This is interpreted as a possible model for differential hormonal release, system disorders and noise amplification in the endocrine system. The modification includes a response of the hypothalamus to short feedback.

1. INTRODUCTION

Earlier systems of differential equations modeling the male mammalian reproductive endocrine system have exhibited a Hopf bifurcation (Smith, 1981). The basic model consisted of a negative feedback system of three ordinary differential equations. The bifurcation parameter is a counterpart of biological age, and it was suggested that the onset of the limit cycle qualitatively mimics the onset of puberty. Experimental data obtained from laboratory animals display noisy almost-periodic time series, however. In a recent paper, a modification was made to one of the feedback functions for such a dynamical system (corresponding in our model

to response of the hypothalamus to testosterone, which we call <u>long</u>
<u>feedback</u>). This produced a chaotic attractor, in the sense of probability
at least, in place of the limit cycle (Rössler, Gotz, and Rössler, 1979;
see also Sparrow, 1981). This could be useful in adapting the model to
better mimic the data, which is characteristically noisy. However, this
modification of the long feedback function (raising the skirt to a V
shape) is difficult to interpret physiologically.

   In this paper, we achieve a similar result with different
modifications of the model. First, the V-shaped modification to the long
feedback function is replaced by a slight kink. The chaotic probable-
attractor persists, as shown in Appendix B. With a second modification,
based on the concepts of complex dynamical systems theory (Abraham, 1982a,
1982b, 1982c), we introduce a response of the hypothalamus to the
pituitary hormone, which we call <u>short</u> <u>feedback</u>. Then, in an extensive
series of simulations, we discover a second periodic attractor, shown in
Appendix C, and a rich bifurcation diagram, indicated in Appendix D.

2. THE BASIC SYSTEM FOR LONG FEEDBACK

Here we review the simple model which exhibits a Hopf bifurcation. In
this model, the endocrine system consists of three hormonal sources: the
hypothalamus (H), the pituitary (P), and the gonads (G). Each of these
emits a single hormone: luteinizing-hormone releasing hormone(R) from H,
luteinizing hormone(L) from P, and testosterone(T) from G. We represent
the appropriate serum concentration of each (normalized or rescaled
values, relative to standard levels) by x, y, z, respectively. The domain
of the basic dynamical system is Euclidean three-space, with these
relative concentrations as coordinates. The system of first-order
equations for the rates of secretion of the three hormones into the
relevant circulatory systems represents the long feedback loop of Figure
1. The basic equations are:

$$\dot{x} = f(z) - x$$
$$\dot{y} = h(x) - y \qquad\qquad (1)$$
$$\dot{z} = g(y) - z$$

Choosing the simplest forms for the three stimulus-response (S-R)
functions f, h, and g (as shown in Fig. 2), a Hopf bifurcation is obtained

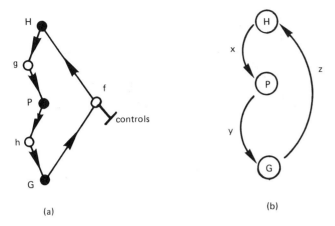

(a)   (b)

FIGURE 1.   THE LONG FEEDBACK CYCLE.
(a) schematic diagram   (b) pictorial diagram

in system (1) by varying the parameter a. We regard this parameter as a
control of the static coupling function, f. The other parameter, f[0], is
held fixed.

The basic system is physiologically more plausible if all three S-R
functions are smooth ramps (Michaelis-Menten functions). For ease of
digital simulation, we shall use piecewise-linear ramps for our improved
basic system, shown in Fig. 3. This improved basic system exhibits the
same qualitative features as the original basic system. The trajectories
are shown in Figs. A1 and A2 in Appendix A. The traditional analysis is
summarized here for review.

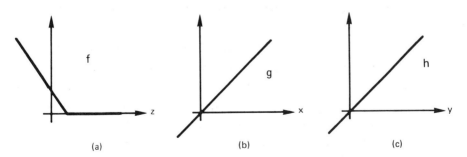

FIGURE 2.   THE SIMPLEST S-R FUNCTIONS OF THE BASIC SYSTEM.

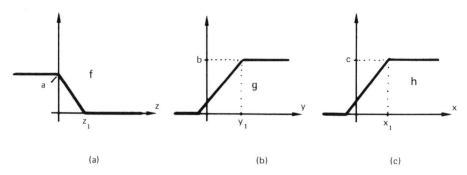

(a)   (b)   (c)

FIGURE 3.   PIECEWISE-LINEAR RAMP S-R FUNCTION OF THE IMPROVED
            BASIC SYSTEM.

    We begin by looking for the critical points of the improved basic
system.   Setting the three rates to zero, we have:

$$x = f(z)$$
$$y = h(x)$$
$$z = g(y)$$
(2)

or, equivalently,

$$x = f(g[h(x)]) = F(x),$$
(3)

where F = fogoh, which we shall call the zero discriminant function.
Zeroes of the vector field are revealed as fixed points of this function,
or equivalently, as intersections of its graph with the diagonal, D =
$\{x, x\}$.

    In the original system (Figure 2) with g and h equal to the identity,
F is equal to f, and is monotone decreasing, as shown in Figure 4a.   In
the improved system (Figure 3), F is still monotone, as shown in Figures
4b and 4c.   The toe of F at $x = x_1$ occurs at the smallest of the three
saturation stimuli, so even in the improved system (Figure 3), all three
cases of Fig. 4 are possible.   In each, however, there is only one
crossing of the diagonal, D, at $x = x_0$.   Thus $F(x_0) = D(x_0) = x_0$, and x_0
determines the unique critical point of the improved system,

$$P_0 = (x_0 \ y_0, \ z_0) = (x_0, \ h[x_0], \ g(h(x_0)))$$
(4)

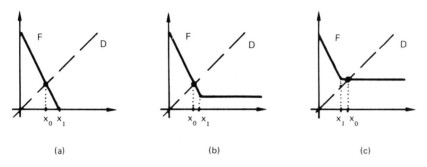

FIGURE 4.   THE ZERO DISCRIMINANT FUNCTION.

Assuming that this point is typical (that is, not at a joint of f, g, or h), then the system is linear at p_0. Or, if we put

$$
\begin{aligned}
X &= x - x_0 \\
Y &= y - y_0 \\
Z &= z - z_0
\end{aligned}
\tag{5}
$$

the system becomes

$$
\begin{aligned}
X' &= -aZ - X \\
Y' &= cX - Y \\
Z' &= bY - Z
\end{aligned}
\tag{6}
$$

or, in matrix notation, $P' = AP$

where

$$
A =
\begin{vmatrix}
-1 & 0 & -a \\
c & -1 & 0 \\
0 & b & -1
\end{vmatrix}
\tag{7}
$$

A has eigenvalues, $(-abc)^{1/3} - 1$, as shown in Figure 5. If $m = 1$, we get a Hopf bifurcation, so we exclude this case (see Fig. 4a(cube root)) and increase the radius, m, via a, b, or c.

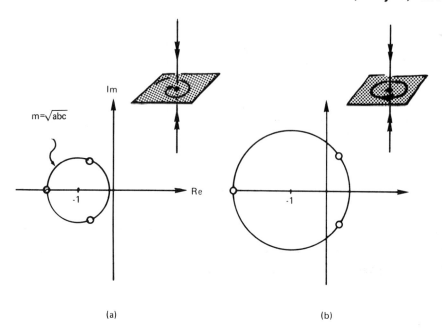

(a)   (b)

FIGURE 5. CHARACTERISTIC EXPONENTS.

The spectrum of A, with corresponding phase portraits.

(a) Rest point, small m   (b) Oscillation, large m

3. THE CHAOTIC ATTRACTOR

The modified long feedback reponse function introduced by Rössler, Gotz,
and Rössler (1979), and used by them and by Sparrow (1981) to obtain a
convergent sequence of bifurcations leading to a chaotic probable-
attractor, is shown in Figure 6(a). The large rise to the right is not
actually needed. We replace this by a piecewise linear form of the
Michaelis-Menton S-R function, as in the basic model of the preceding
section. This is shown in Figure 6(b). With the height of the skirt as a
control parameter, the sequence of bifurcations still converges to a
chaotic attractor. The details are shown in Appendix B. It may yet be
unacceptable physiolgically but monotonicity is violated only in a very
small region of hormone concentrations.

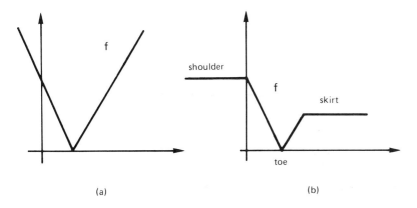

(a)   (b)

FIGURE 6.   THE LONG FEEDBACK FUNCTIONS FOR CHAOS.

(a)   The original V-function.   (b)   The modified function.

4.   THE NEW MODEL WITH SHORT FEEDBACK

We now include short feedback between the hypothalamus and the pituitary, by supposing the hypothalamus (H) is sensitive to L, emitted by the pituitary (P), as well as to T from the gonads (G). Thus we replace the scheme of Figure 1 by the modified one of Figure 7, and add the new

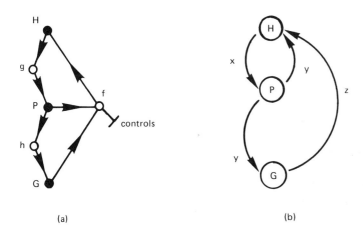

(a)   (b)

FIGURE 7.   THE MODIFIED SCHEME WITH SHORT FEEDBACK.

(a)   Schematic   (b)   Pictorial

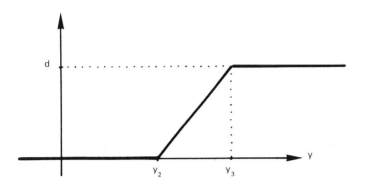

FIGURE 8.   THE SHORT FEEDBACK FUNCTION.

(a)

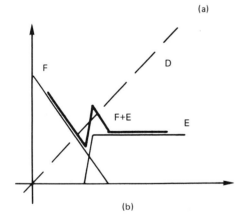

(b)

FIGURE 9.   ZEROS OF THE MODIFIED SYSTEM.

(a)   Choice of constants   (b)   Zero discriminant

response function, e(y) to the x equation. We assume the new S-R
function, e, is a piecewise-linear ramp, like f, g, and h, as shown in
Figure 8, obtaining the modified system,

$$\dot{x} = e(y) + f(z) - x$$
$$\dot{y} = h(x) - y \qquad\qquad\qquad (8)$$
$$\dot{z} = g(y) - z$$

Seeking critical points as before, we see that (x_0, y_0, z_0) is a
critical point of the system (8) only if $y_0 = h(x_0)$, $z_0 = g(y_0)$, and $x_0 = G(x_0)$, where:

$$G(x) = f(g\,[h(x)]) + e(h[x]) = F(x) + E(x)$$

Unlike that for F previously, G is no longer monotone. We will
suppose the parameters of the functions are related as shown in Figure 9a.
Then the graphs of E and F are as in Figure 9b. The toe joint, y_2,
shoulder joint, y_3, and height, d, of the new sensitivity function, e, now
join the height, a, and toe, b, of the original function, f, as the
control parameters of the static coupling function, f+e.

5. THE TWO PERIODIC ATTRACTORS

We first fix the function f as one which provides a periodic attractor in
the basic model of Section 2, as shown in Fig. A2. We next choose
appropriate values for the toe and shoulder joints of the new response
function, e. Then, regarding the height of e as the sole control
parameter of the coupling function to the hypothalamus, simulation reveals
the bifurcation diagram of the double fold catastrophe. A similar result
has been discovered in a sequence of enzymatic reactions (Decroly and
Goldbeter, 1982).

The actual trajectories are shown in the sequence of computer
drawings in Appendix C. This diagram has been described in great detail
as a model for intermittency and noise amplification (Abraham, 1983a).
Here is the idea. Reducing the three-dimensional state space of the
endocrine system fictitiously to one (for example, by observing only the
amplitude of the oscillations) we may portray this bifurcation diagram in
a plane, as shown in Fig. 10.

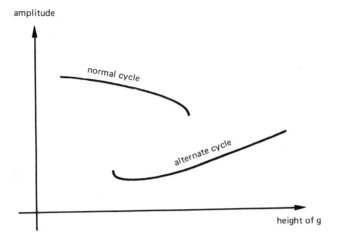

amplitude

height of g

FIGURE 10. PERIODIC DOUBLE FOLD CATASTROPHE.

Increasing the control parameter from the far left across both bifurcation points, we observe a catastrophic change from the normal cycle to the alternate cycle at the right bifurcation point. Returning the control to its original value, we observe a catastrophic return to the normal cycle at the left bifurcation point. This is a hysteresis loop in the control space. If, for example, the control parameter were determined by another dynamical system, we might observe changes between the two cycles intermittently. This provides a model for intermittency in the context of serially coupled dynamical systems. In this particular case, both of the states intermittently occupied are periodic.

Now suppose that the scale of the amplitude (vertical) axis in Figure 10, relative to the control (horizontal) scale, is small. Thus, a very small oscillation in the control parameter results in a relatively large alternation between the normal and alternate cycles. Further, a small but noisy variation of the control parameter crossing both bifurcation points repeatedly results in a large and noisy variation between the two cycles. This is a model for noise amplification in the context of serially coupled dynamical systems. We see in Appendix C that in our modified model for the endocrine system, this exaggerated scale relationship applies. Small changes in the control parameter (height of the skirt between 0.70 and 0.95) produce a large, hysterical variation between the normal cycle and the alternate state. The magnitude of this variation is clearly shown in Figure C2(e).

6. APPLICATIONS TO THE PHYSIOLOGICAL SYSTEM

Suppose that in Fig. 10, the upper attractrix corresponds to the normal cycle exhibited in adults. The lower attractrix is an oscillation of smaller amplitude, and totally different period. This suggests three different possible physiological implications.

First, it is known that the hormone R effects the release of another hormone, follicle-stimulating hormone, from the pituitary in addition to L. The two different cycles of the model may correspond to the required stimuli for the release of the different hormones. The switching between the two release mechanisms would be effected by variations in the control parameter. This may thus be a mechanism for differential release of the different hormones.

Second, in view of the results of Appendix C, we see that amplification by the endocrine system of noise in the control parameter (sensitivity of H to R) could account for observed experimental noise in the serum concentration data of L and G.

As a third possible implication, suppose that the parameters in the model correspond to two different cycles, one characterizing the normal state, and the other a pathological state corresponding to some disorder. Then, based on the diagram, we may propose two therapeutic strategies. First, as dynamical systems latch on attractors, we may try to force the system from one attractor into the basin of the other, where it will then latch. Although catastrophic, this seems to have found some support in recent clinical findings (Jaffe, 1982). On the other hand, if a way were known to adjust the height parameter of the short feedback response function, e, of the hypothalamus to luteinizing harmonic, then a small, gradual, temporary decrease would achieve the same effect.

7. CUSP CATASTROPHES

We fix the basic response function, f, as above. Fixing the height of the new response function, e, at a convenient value, we now vary the toe and shoulder joints to the right and left. The effect of moving the toe is to raise and lower the lower spike of the zero discriminant along the incline of f, as shown in Figure D1(d), in Appendix D. As this spike passes through the diagonal, D, the two critical points annihilate. The periodic attractor, originally created from one of these by a Hopf bifurcation, eventually disappears (becomes nonattractive) as well, as shown in Fig. D1(e).

Similarly, moving the shoulder joint of e to the right lowers the upper spike along the incline parallel to f, as shown in Fig. D2(d). When this spike passes through the diagonal, D, another static annihilation catastrophe occurs. Taking these events together, we see that we have a static cusp catastrophe, with toe and shoulder joint hormonal concentrations as control parameters.

If the slope of f were above the Hopf bifurcation value (-8), this cusp catastrophe would involves two point attractors and a saddle point, the static cusp catastrophe. If the slope of f is below the Hopf bifurcation value but close to it (which it was not, in our simulations) then the two periodic attractors and the periodic saddle cycle would also be related in a cusp catastrophe, the periodic cusp catastrophe. Likewise, if the skirt of f were lifted as in Section 3 above, we conjecture that three chaotic limit sets (two attractors and a saddle) would be related in a chaotic cusp catastrophe. We have not verified this behavior with simulations, but the computer drawings of Appendix D are highly suggestive.

8. CONCLUSIONS

In the literature of complex dynamics (Abraham, 1982, 1983) it is proposed that serial networks of dynamical schemes provide a useful strategy for the architect of dynamical models and applications. Further, it is suggested that serial chains are important cases, and serial cycles are most important. The basic model of Smith (1981) was chosen as a test case for the theory. Here, we have put these proposals to a practical test. This results in a plausible modification to the basic model, a serial cycle with three dynamic nodes (Fig. 1), through the addition of one edge (Fig. 7). Simulations of the resulting complex dynamical scheme, exploring the effects of variations in its five parameters, reveal a rich bifurcation diagram. The qualitative interpretation of this diagram may enable better modeling and simulation of endocrine systems, and the discovery of new therapeutic strategies. Further, the application of this style of modeling to other kinds of complex systems may likewise create models for them, in which the basic phenomena of complex dynamics (chaos, intermittency, catastrophes, hysteresis, and so on) may be fitted by the model.

APPENDICES: COMPUTER DRAWN ORBITS

These are computer plots of two-dimensional projections of trajectories of
the dynamical models into the (x,z) plane. They have been computed on a
DEC VAX 11/780 with 2.5MB of main memory, and a floating point
accelerator.   The ORBIT program, written in C and run under UNIX
V7/4.1BSD, uses a fourth-order Runge-Kutta algorithm with Richardson
extrapolation. The output was viewed on a Ramtek 610 color graphics
terminal (320×240 pixel resolution) and plotted on a Tektronix plotter.

A. BASIC MODEL, HOPF BIFURCATIONS.

Fig. A1. BASIC MODEL, STATIC DOMAIN.
The normal point attractor.
(a) Long feedback function.
(b) Short feedback function.
(c) Zero discriminant and diagonal.
(d) Detail of the intersection.
(e) Trajectories.

Fig. A2. BASIC MODEL, PERIODIC DOMAIN.
The normal periodic attractor.
(a) Long feedback function.
(b) Short feedback function.
(c) Zero discriminant and diagonal.
(d) Detail of the intersection.
(e) Trajectories.

B. RAISED SKIRT MODEL, ONSET OF CHAOS.

Fig. B1. RAISED SKIRT MODEL, PERIODIC DOMAIN.
The perturbed periodic attractor.
(a) Long feedback function.
(b) Short feedback function.
(c) Zero discriminant and diagonal.
(d) Detail of the intersection.
(e) Trajectories

Fig. B2. RAISED SKIRT MODEL, TRIPLE-PERIODIC DOMAIN.
Attractive cycle exhibiting triple the normal period.
(a) Long feedback function.
(b) Short feedback function.
(c) Zero discriminant and diagonal.
(d) Detail of the intersection.
(e) Trajectories.

Fig. B3. RAISED SKIRT MODEL, CHAOTIC DOMAIN.
The chaotic attractor of Rössler et al.
(a) Long Feedback function.
(b) Short feedback function.
(c) Zero discriminant and diagonal.
(d) Detail of the intersection.
(e) Trajectories.

C. SHORT FEEDBACK MODEL, BIRHYTHMICITY.

Fig. C1. SHORT FEEDBACK MODEL, BIMODAL DOMAIN.
The normal periodic attractor dominates, but a new point
attractor has been born.
(a) Long feedback function.
(b) Short feedback function.
(c) Zero discriminant and diagonal.
(d) Detail of the intersection.
(e) Trajectories.

Fig. C2. SHORT FEEDBACK MODEL, BIRHYTHMIC DOMAIN
The new periodic attractor, inside the normal cycle, has appeared
after a Hopf bifurcation of the new point attractor.
(a) Long feedback function.
(b) Short feedback function.
(c) Zero discriminant and diagonal.
(d) Detail of the intersection.
(e) Trajectories.

Fig. C3. SHORT FEEDBACK MODEL, ALTERNATE-PERIODIC DOMAIN.
The normal periodic attractor has destabilized, but the alternate
periodic attractor remains.

(a) Long feedback function.

(b) Short feedback function.

(c) Zero discriminant and diagonal.

(d) Detail of the intersection.

(e) Trajectories.

Fig. C4. SHORT FEEDBACK MODEL, ALTERNATE-STATIC DOMAIN.

The alternate periodic attractor has become a point attractor, through an inverse Hopf bifucation.

(a) Long feedback function.

(b) Short feedback function.

(c) Zero discriminant and diagonal.

(d) Detail of the intersection.

(e) Trajectories.

D. SHORT FEEDBACK MODEL, CUSP CATASTROPHES.

Fig. D1. SHORT FEEDBACK MODEL, VARIATION OF THE TOE PARAMETER.

Compare with Fig. C2(e). Moving the toe to the left moves the lower spike upwards, along the incline of f, as shown in Fig. D1(d) here. In this case, the normal cycle has suffered a periodic annihilation catastrophe, involving a collision with its separator.

(a) Long feedback function.

(b) Short feedback function.

(c) Zero discriminant and diagonal.

(d) Detail of the intersection.

(e) Trajectories.

Fig. D2. SHORT FEEDBACK MODEL, VARIATION OF THE SHOULDER PARAMETER.

Compare with Fig. C2. Moving the shoulder to the right lowers the upper spike along the incline parallel to f, as shown in (d) here. In this case, the alternate limit cycle has become a perodic repellor.

(a) Long feedback function.

(b) Short feedback function.

(c) Zero discriminant and diagonal.

(d) Detail of the intersection.

(e) Trajectories.

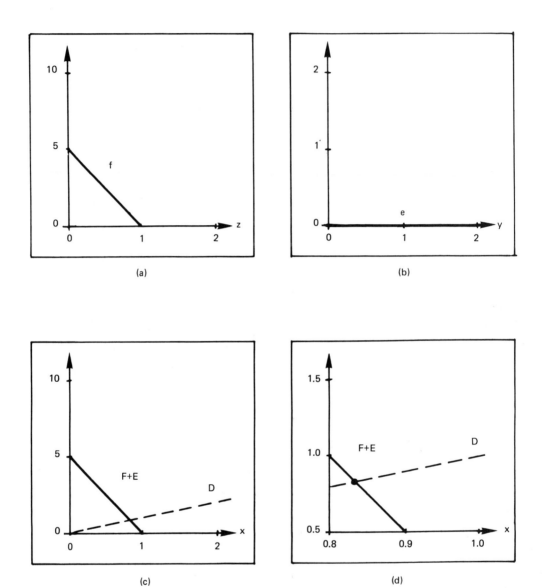

Fig. A1. BASIC MODEL, STATIC DOMAIN.

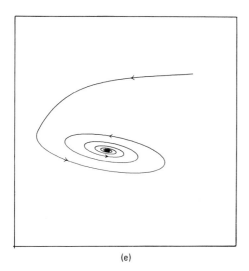

(e)

Fig. A1. BASIC MODEL, STATIC DOMAIN.

The normal point attractor.

(a) Long feedback function.

(b) Short feedback function.

(c) Zero discriminant and diagonal.

(d) Detail of the intersection.

(e) Trajectories.

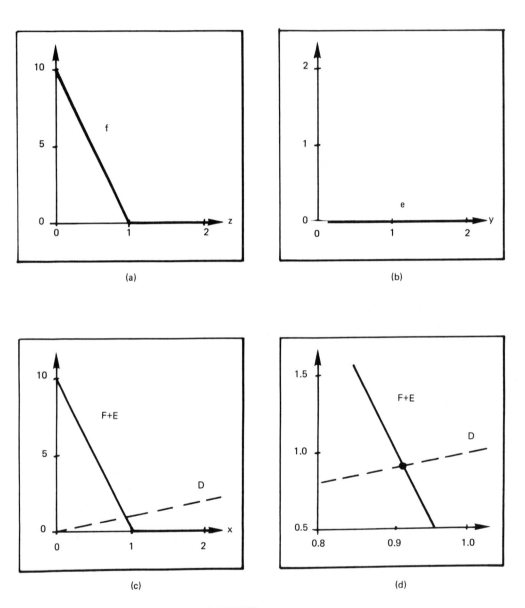

Fig. A2. BASIC MODEL, PERIODIC DOMAIN.

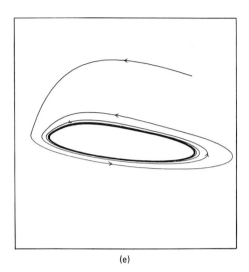

(e)

Fig. **A**2. BASIC MODEL, PERIODIC DOMAIN.

The normal periodic attractor.

(a) Long feedback function.

(b) Short feedback function.

(c) Zero discriminant and diagonal.

(d) Detail of the intersection.

(e) Trajectories.

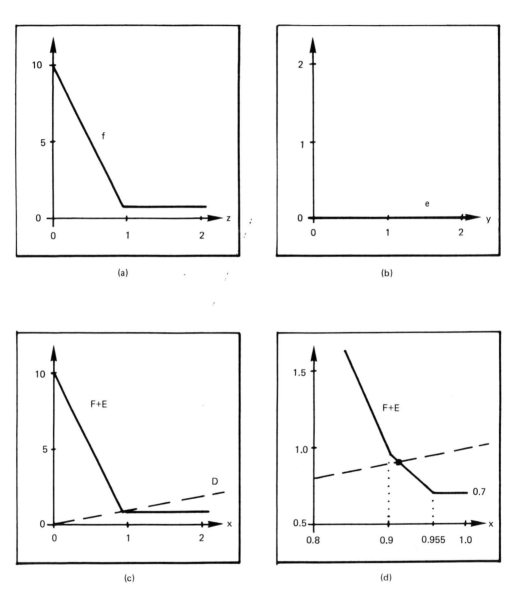

Fig. B1. RAISED SKIRT MODEL, PERIODIC DOMAIN.

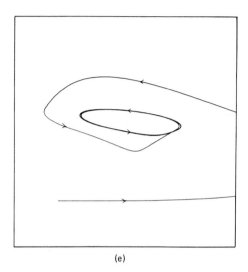

(e)

Fig. B1. RAISED SKIRT MODEL, PERIODIC DOMAIN.
The perturbed periodic attractor.

(a) Long feedback function.

(b) Short feedback function.

(c) Zero discriminant and diagonal.

(d) Detail of the intersection.

(e) Trajectories.

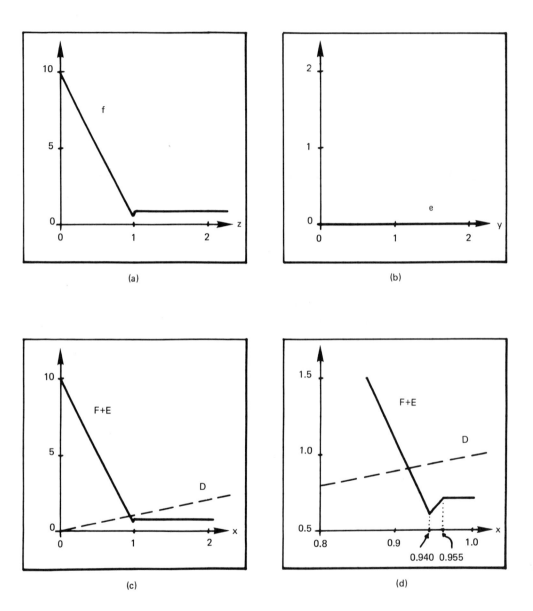

(a)

(b)

(c)

(d)

Fig. B2.   RAISED SKIRT MODEL, TRIPLE-PERIODIC DOMAIN.

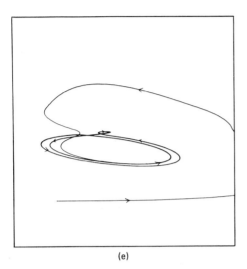

(e)

Fig. B2. RAISED SKIRT MODEL, TRIPLE-PERIODIC DOMAIN.
Attractive cycle exhibiting triple the normal period.

(a) Long feedback function.

(b) Short feedback function.

(c) Zero discriminant and diagonal.

(d) Detail of the intersection.

(e) Trajectories.

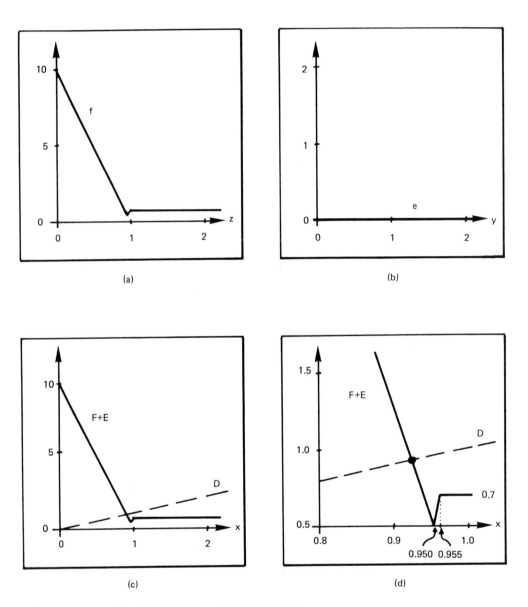

Fig. B3.   RAISED SKIRT MODEL, CHAOTIC DOMAIN.

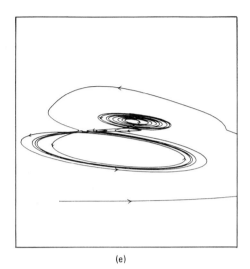

(e)

Fig. **B3**. RAISED SKIRT MODEL, CHAOTIC DOMAIN.

The chaotic attractor of Rössler et al.

(a) Long Feedback function.

(b) Short feedback function.

(c) Zero discriminant and diagonal.

(d) Detail of the intersection.

(e) Trajectories.

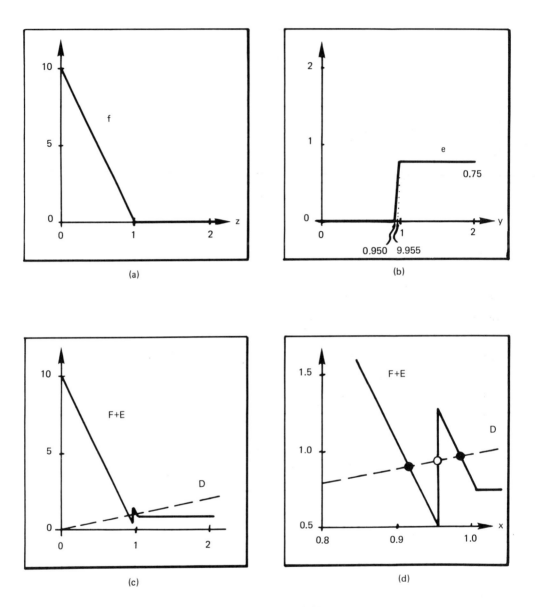

Fig. C1.   SHORT FEEDBACK MODEL, BIMODAL DOMAIN.

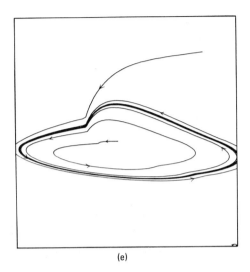

(e)

Fig. C1. SHORT FEEDBACK MODEL, BIMODAL DOMAIN.

The normal periodic attractor dominates, but a new point attractor has been born.

(a) Long feedback function.

(b) Short feedback function.

(c) Zero discriminant and diagonal.

(d) Detail of the intersection.

(e) Trajectories.

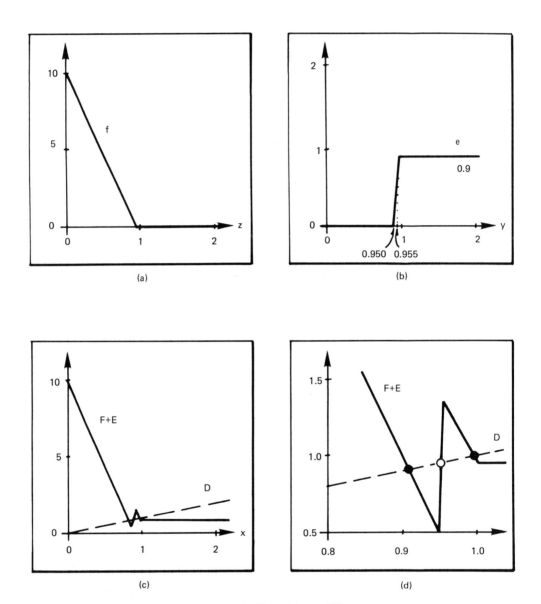

Fig. C2. SHORT FEEDBACK MODEL, BIRHYTHMIC DOMAIN

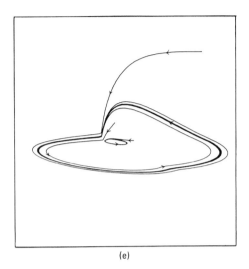

(e)

Fig. C2. SHORT FEEDBACK MODEL, BIRHYTHMIC DOMAIN

The new periodic attractor, inside the normal cycle, has appeared
after a Hopf bifurcation of the new point attractor.

(a) Long feedback function.

(b) Short feedback function.

(c) Zero discriminant and diagonal.

(d) Detail of the intersection.

(e) Trajectories.

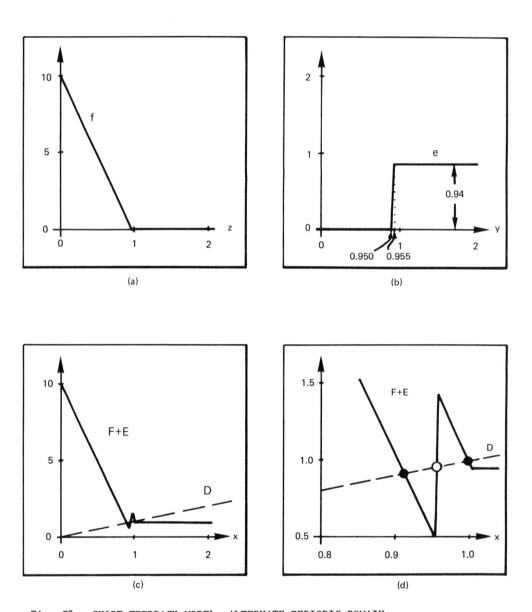

Fig. C3. SHORT FEEDBACK MODEL, ALTERNATE-PERIODIC DOMAIN.

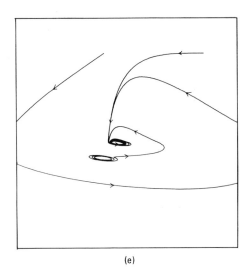

(e)

Fig. C3. SHORT FEEDBACK MODEL, ALTERNATE-PERIODIC DOMAIN.
The normal periodic attractor has destabilized, but the alternate
periodic attractor remains.

(a) Long feedback function.

(b) Short feedback function.

(c) Zero discriminant and diagonal.

(d) Detail of the intersection.

(e) Trajectories.

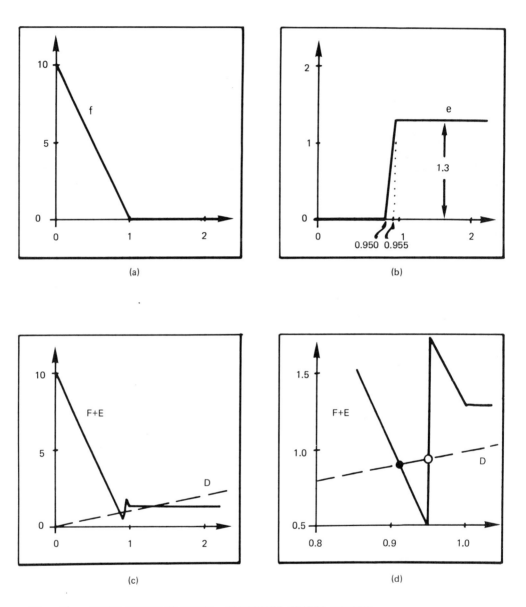

Fig. C4. SHORT FEEDBACK MODEL, ALTERNATE-STATIC DOMAIN.

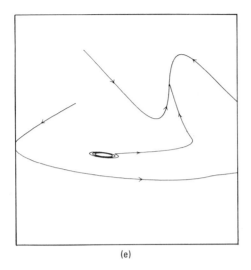

(e)

Fig. C4. SHORT FEEDBACK MODEL, ALTERNATE-STATIC DOMAIN.
The alternate periodic attractor has become a point attractor,
through an inverse Hopf bifucation.
(a) Long feedback function.
(b) Short feedback function.
(c) Zero discriminant and diagonal.
(d) Detail of the intersection.
(e) Trajectories.

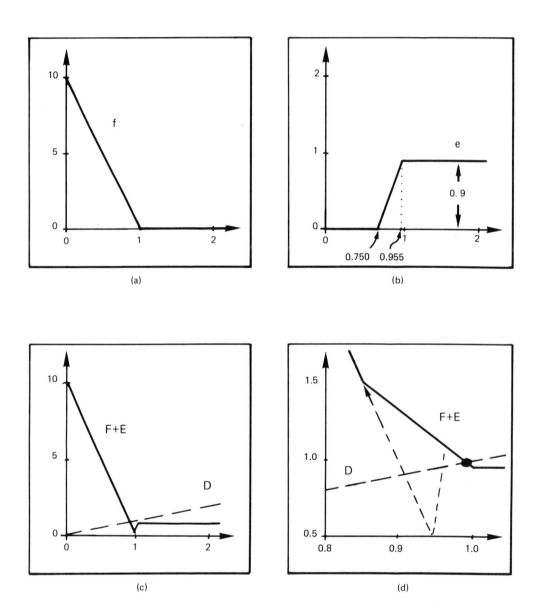

Fig. D1. SHORT FEEDBACK MODEL, VARIATION OF THE TOE PARAMETER.

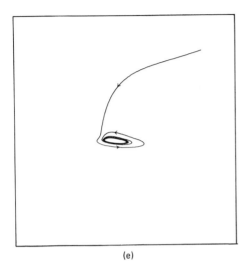

(e)

Fig. D1. SHORT FEEDBACK MODEL, VARIATION OF THE TOE PARAMETER.
Compare with Fig. C2(e). Moving the toe to the left moves the
lower spike upwards, along the incline of f, as shown in Fig.
D1(d) here. In this case, the normal cycle has suffered a
periodic annihilation catastrophe, involving a collision with its
separator.

(a) Long feedback function.

(b) Short feedback function.

(c) Zero discriminant and diagonal.

(d) Detail of the intersection.

(e) Trajectories.

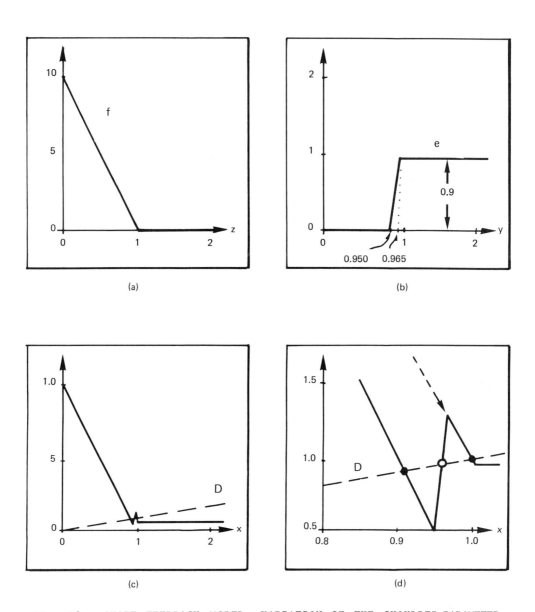

Fig. D2. SHORT FEEDBACK MODEL, VARIATION OF THE SHOULDER PARAMETER.

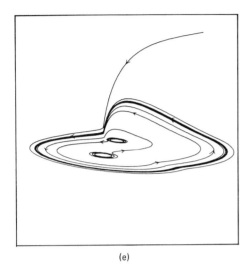

(e)

Fig. D2.   SHORT FEEDBACK MODEL, VARIATION OF THE SHOULDER
PARAMETER.
Compare with Fig. C2.   Moving the shoulder to the right lowers
the upper spike along the incline parallel to f, as shown in (d)
here.   In this case, the alternate limit cycle has become a
perodic repellor.
(a) Long feedback function.
(b) Short feedback function.
(c) Zero discriminant and diagonal.
(d) Detail of the intersection.
(e) Trajectories.

ACKNOWLEDGEMENTS

This work was born in Guelph, and clearly owes much to Otto Rössler at the
conference in March, 1981. The modified system resulted from further
discussions in Guelph and Tübingen. The computer simulations were done in
Santa Cruz, using the ORBIT software created by Ralph Abraham and Tod
Blume, and adapted to this problem with the help of Bob Lansdon. We are
grateful to Arthur Fischer, Otto Rössler and Alan Garfinkel for helpful
conversations, and to the Universities of California, Guelph, and Tübingen
for their support and hospitality during this project.

BIBLIOGRAPHY

Abraham, R. H., 1983a. Categories of dynamical models, in: T. M. Rassias
(ed.), Global Analysis-Analysis on Manifolds, Teubner, Leibzig (in press).

Abraham, R. H., 1983b. Dynamical models for physiology, preprint.

Abraham, R. H., 1983c. Chaostrophes, Intermittency, and Noise, this
volume.

Abraham, R. H. and C. D. Shaw, 1982. Dynamics, the Geometry of Behavior,
Part 1: Periodic Behavior, Aerial Press, Box 1360, Santa Cruz, CA 95061.

Abraham, R. H. and C. D. Shaw, 1983. Dynamics, a visual introduction, in:
F. E. Yates (ed.), Self-Organizing Systems, Plenum, (to appear).

Decroly, A. and A. Goldbeter, 1982. Birhythmicity, chaos, and other
patterns of temporal self-organization in a multiply regulated biochemical
system, Proc. Natl. Acad. Sci. USA 79: 6917-6921.

Jaffe, R. B., 1982. Pump that helps women ovulate, San Francisco
Chronicle, p. 4, Sept. 10, 1982.

Rössler, R., R. G. Gotz, and O. E. Rössler, 1979. Chaos in endocrinology,
Biophys. J. 25: 216(a).

Smith, W. R., 1981. Hypothalamic regulation of pituitary secretion of
luteinizing hormone. II. Feedback control of gonadotropin secretion.
Bull. Math. Biol. 42:57.

Sparrow, C. T., 1981. Chaos in a three-dimensional single loop feedback
system with a piecewise linear feedback function, J. Math. Anal. Appl.
83: 275-291.

4

An Index for Chaotic Solutions in Cooperative Peeling

Okan Gurel

IBM Cambridge Scientific Center
Cambridge, Massachusetts

Among the most important solutions of a dynamical system are those which
remain in a bounded subset of the solution space. Based on the behavior
of characteristic flows at singular points in the solution space, the
concepts of the functional dimension (f-dimension) and (based on this
dimension) an index for the solutions bifurcating from these singular
points are defined. Chaotic solutions may be characterized by referring
to this index. Some preliminary notions leading to these definitions are
given and the concepts are illustrated by examples.

0. INTRODUCTION

Among the interesting solutions of a dynamical system are the ones
which remain in a bounded subset of the solution space. The
singular solution where the right hand sides of the differential (system)
equations, $dx/dt=f(x)$, vanish is the most fundamental and the most
elaborated one in the relevant literature. Once at the singular solution,
the system remains there as $t \to \infty$. Similarly, limit cycles have been
recognized and incorporated into studies of physical systems as solutions
remaining in bounded subsets of the solution space.

71

Although predicted and partially studied in the past, the appearance of "attractors", "chaotic objects", etc. has increased recently in the literature, and examples of simple systems of low dimension, $N \leq 3$, have been constructed.

There are specific ways to characterize the early objects. Poincaré's observations on the singular solutions and their classification as focus, center and saddle point, and the subsequent incorporation of the stability properties of these "local" solutions as characterizing the "global" behaviour of the system possessing such points are well known advances in the field. The extension of the notion of stability to limit cycles is a natural development since in nonlinear systems, even in two-dimensional ones, limit cycles enter almost as commonly as singular points appear in linear systems.

The difficulty with such "characteristic" objects in higher dimensions, $N \geq 3$, is due to many factors. However, it is only recently that we have realized the relationship between such "new" objects and known "old" physical phenomena, and thus we are forced to consider these objects seriously to better understand their behavior. This requires modifying our already established thinking, if not inventing new ways to interpret such discoveries. The entire field of turbulence and methods of unravelling the behavior of oscillations with multiple periods via one dimensional mapping techniques are good examples of this recent trend.

In this spirit we describe how we can characterize such solutions. We summarize certain recent attempts. Some of these solutions appear to behave "stochastically". Therefore physicists use stochasticity as part of the analysis of such objects. The view presented here emphasizes and is based on "deterministic", or perhaps more correctly "structural", interpretations of the system rather than "probabilistic" ones. This is a crucial issue and the merits of one or the other approach should be viewed without bias by the reader.

I. DEFINITION OF AN INDEX

Before we define the stability index we first present some preliminary concepts. These are characteristic solutions, related stability concepts and the recently defined notion of f-dimension.

1. Characteristic Solutions

A dynamical system in E^n is described by

$$f_p: X \rightarrow X, \quad X \subset E^n$$

where the subscript p refers to the dependence of f on the parameters. Considering a dynamical system where f_p is taken as the right-hand sides of differential equations, one can define characteristic solutions of the dynamical system in terms of f_p. The usual definition is then stated as:

Definition 1a. Singular Solution: is a point x_o such that $f_p(x_o) = 0$.

2. Stability of Solutions [1,2,3]

The stability of a given singular solution can be discussed by referring to the flows approaching and leaving the solution. The existence and the number of these flows can easily be determined. We first linearize the equations about the given singular point, determine the eigenvalues, and represent the results in terms of w_{m1} and w_{m2} corresponding to approaching and leaving flows. In the rest of this paper, the following notations will be used:

Linearized system:	Lf_p at x_o
Characteristic equation:	$\left| Lf_p - s \right| = 0$
Eigenvalues:	$s_1, \dots, s_n$
Half-flows:	$1/2 w_{m1}$ (stable) and $1/2 w_{m2}$ (unstable)

For each real eigenvalue which is negative(positive) a pair of stable (unstable) half-flows are denoted by $1/2\ w_{m1}$ ($1/2\ w_{m2}$).

The Definition 1a is restated in terms of flows as

Definition 1b. A singular point x_o is a singular solution having n-k stable flows w_{m1} and k unstable flows w_{m2}, with k=0 implying a stable singular point.

Definition 2. An exploded point x_e is a solution having only one stable (or unstable) flow w_{m1} (or w_{m2}) approaching (or leaving) it.

Definition 3. A limit cycle L is a set of points in X which is the image of a solution satisfying f_p: L → L having only two stable (or unstable) flows w_{m1} (or w_{m2}), approaching (and/or leaving) it, respectively.

Remark:

The simplest limit cycles are obtained in a two-dimensional space, although in three and higher dimensions one can still obtain them. These

may exhibit multiple periodicities in multiple coordinates. Limit cycles
with double period are the simplest of these. A periodic solution on T^3
is an example. The generic name for these limit cycles was proposed in
[3] as limit bundles. x_e does not satisfy f = 0; thus it is not
stationary but oscillating, and without any period. On the other hand, a
limit bundle is a limit cycle with multiple periods.

3. f-dimension [2]

Definition 4.   The   f-dimension (functional dimension) of a solution x is
one less than d, the sum of the number of stable and unstable flows
approaching (and/or) leaving it:

$$\text{f-dim } \mathbf{x} = d-1.$$

The f-dimension and other properties of the singular solutions discussed
above are given in the table below.

4. A Stability Index [2]

Definition 5.   The stability index of a singular point is defined as the
difference between the number of stable flows w_{m1}, (n-k) (i.e., the
dimension of the stable manifold), and the number of unstable flows w_{m2},
(i.e., the dimension of the unstable manifold,). Hence

$$I_s = (n-k) - k = n - 2k$$

Type of Singular Solution	f-dim	Minimum Dimension of Space	Number of Independent Periods
Singular Point, x_o	n-1	1	0
Exploded Point, x_e	0	1	0
Limit Cycle, L	1	2	1
Limit Bundle, LB	1	≥ 3	≥ 2

II. CONCEPT OF COOPERATIVE PEELING

In the literature, the bifurcation of one singular point (or a periodic solution) has been discussed. The concept that we would like to use in illustrating the stability index is the cooperative (simultaneous) peeling of multiple singular solutions [4]. The method of using the half-flows to determine the stability of the resulting object may be applied to such complex behavior.

1. Cooperative Peeling of Multiple Solutions [4]

One of Poincaré's necessary conditions for bifurcation at a singular point $x_0(p_0)$ is that the determinant of M evaluated at x_0 vanishes at the parameter value, p_0. Thus denoting

$$\Delta(x,p) = |L(x,p)|$$

and $\Delta_c(x,p)$ is the determinant of a diagonal matrix corresponding to M which satisfies

$$\Delta_c(x_0,p) = m_{11}(x_0,p) \cdots m_{nn}(x_0,p)$$

The critical limiting set CLS(X) contains all the characteristic solutions. Let us assume that there are k singular points $x_1, x_2, \ldots x_k$ in addition to x_0. Thus we have the matrices L_0, L_c^1, $\ldots$, L_c^k. These are combined into a matrix L*, which will be used to indicate the cooperative behavior. The elements of the combined matrix are

$$M* = \begin{vmatrix} m_{11}^{0} & m_{11}^{1} & \cdots & m_{11}^{k} \\ m_{22}^{0} & m_{22}^{1} & \cdots & m_{22}^{k} \\ \cdot & \cdot & & \cdot \\ \cdot & \cdot & & \cdot \\ m_{nn}^{0} & m_{nn}^{1} & \cdots & m_{nn}^{k} \end{vmatrix}$$

In the case k not equal to n-1, dummy rows or dummy columns of some constant value may be assigned to make the combined matrix a square matrix.

Cooperative Peeling (necessary) Condition I.

> In the case when the determinants of all the individual
> matrices do not vanish, "cooperative peeling" can occur only
> if the determinant of L* vanishes.

The two other conditions of Poincaré, the stability change and change in
the number of solutions are combined in the following condition:

Cooperative Peeling (sufficient) Condition II.

> For cooperative peeling to take place, in addition to
> Condition I it is necessary that CLS(X) is replaced by
> CLS´(X).

Here CLS and CLS´ are the two different sets of solutions before and after
bifurcation (peeling).

2. **Independent Peeling of Multiple Solutions** [5]

Contrary to the phenomenon of cooperative peeling, it is also
possible that multiple singular points may individually (noncooperatively)
peel. In this case we treat each singular point as if the other points do
not exist. Thus, an important characteristic of this is the hyperbolic or
elliptic neighborhood of the individually peeling point.

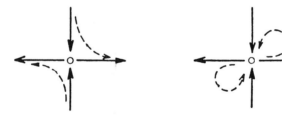

Hyperbolic neighborhood   Elliptic neighborhood

III. OTHER CLASSES OF PEELING

1. **Peeling into Noncritical Limiting Set** [8]

The concept of peeling of an object to yield either a critical
limiting set or a noncritical set is important. It is possible that when
peeling of a stable object takes place, all of the half-flows changing
direction as the stability of the object changes may either all go to a
critical limiting set (CLS) of the system determining the global (overall)

stability of the system, or to a noncritical limiting set (NCLS) which does not belong to the CLS.

2. Combined Peeling in Systems with Multiple Parameters

In this section we discuss peeling of an individual singular point at bifurcation values of different parameters (independently). We need certain definitions.

a. Cardinality [6]

The cardinality of a solution set $\{x_0\}$, Card $\{x_0\}$ is the number of distinct "characteristic" solutions in the set. $\{x_0\}$ splits into two parts, $\{NOx_0\} \cup \{Ox_0\}$ respectively the nonoscillating and oscillating solutions. Thus,

$$\text{Card } \{x_0\} = \text{Card } \{NOx_0\} + \text{Card } \{Ox_0\}$$

The Cardinality Rule:

Card $\{x_0\}$ > Card $\{NOx_0\}$ implies the existence of $\{Ox_0\}$.

b. Global Stability [6]

The global stability of a system is determined by considering the half-flows of the CLS as they are sufficiently far away from the CLS. For example, if the CLS is a stable singular point (the generating singular point), then after the peeling, the CLS (which may or may not include the generating singular point) should have all pairs of half-flows approaching the CLS.

The global stability theorem states

$$\text{Stab}\{x_0\}_{\text{before peeling}} = \text{Stab}\{x_0\}_{\text{after peeling}}$$

c. Dimensionality [6]

The dimensionality of a single solution, either NOx_0 or Ox_0, is defined in terms of the f-dimension. For example, in an n-dimensional system, dim-X = n, a fixed point has the f-dimension as given by definition 4. However, an oscillating solution, Ox_0 may have a range of f-dimension between 0 and dimX-1. This may be stated as:

The Dimensionality Rule:

$$\text{f-dim } (Ox_0) = 0, 1, \ldots, \text{dimX-1}.$$

d. Hyperbolicity [6]

The hyperbolicity of a solution reflects the existence of both stable and unstable flows approaching and leaving that particular solution. Namely, hyperbolicity refers to a stability property of solutions. Based on this definition of hyperbolicity, a rule can be stated to determine the dimension of an oscillating solution (Ox_o) resulting from the bifurcation of a stable NOx_o. A rule combining the stability and f-dimension concepts can then be stated as:

The Hyperbolicity Rule:

(i)   If Stab (NOx_o) = dimX . w_{m2} (unstable), then for the newly created Ox_o

$$f-dim(Ox_o) = dimX - 1.$$

(ii) If Stab (NOx_o) = (dimX-1-k).w_{m1} + (k+1)w_{m2} (hyperbolic), then Stab (Ox_o)=(k+1)w_{m1}, and thus f-dim(Ox_o) = k.

IV.   ILLUSTRATIVE EXAMPLES

The idea of the stability index of a solution of a dynamical system is introduced. This index provides a measure of the stability property as well as the geometry of a solution. Coupled with the concept of cooperative peeling, one can characterize certain solutions as they appear in the process of bifurcations. Here we give some examples to illustrate applications of this approach. These examples are:

1. An exploded point in a cooperative peeling (Lorenz model).

2. An exploded point in an individual peeling (Rössler model).

3. A limit bundle as a noncritical limiting set (Toroidal model).

4. Limit cycles as a result of decomposition in the parameter space (Combined peeling model).

Example 1. Lorenz Attractor (Cooperative Peeling)

Bifurcation of Lorenz System: [7, pp. 14-20]

$dx/dt = -mx + my$

$dy/dt = -x(z-r)-y$

$dz = xy - bz$

For $b > 0$,

1. $r < 1$ $CLS(x) = (S_0$, stable focus)

2. $r > 1$ $CLS(x) = (S_0$, saddle;

S_1, S_2, stable foci)

3. $r > r_c$ $CLS'(x) = (S_0$, saddle,

S_1, S_2, saddle-foci,

LA, Lorenz attractor)

where $r_c = m(A+2)/(2m-A)$, and $A = (b(r-1))^{1/2}$.

Cooperative Peeling: [4, pp. 41-42]

$CLS(x) = (S_0, S_1, S_2)$

$$L^* = \begin{vmatrix} -m & -m & -m \\ (r-1) & -2 & -2 \\ -b & -A^2 & -A^2 \end{vmatrix}$$

Here $\Delta = 0$ for all b and r. The bifurcation (peeling) of $CLS(x)$ into $CLS'(x)$ is cooperative, and results in, (Fig. 1)

$CLS'(x) = (S_0, S_1, S_2, LA)$

The Lorenz attractor can be shown to be an exploded point, (Fig. 2). Moreover the number of possibilities, N_n, yielding this type of exploded point can be determined by the formula [2]

$$N_n = (K_n/2)n!$$

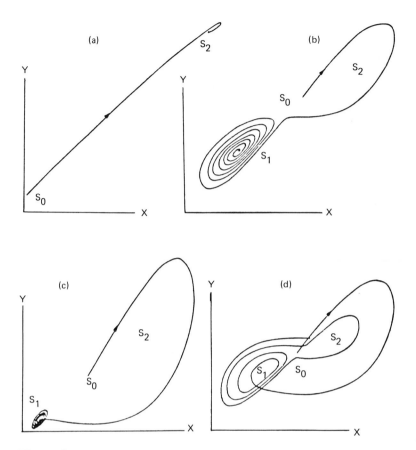

Figure 1.   Cooperative peeling to the Lorenz attractor.

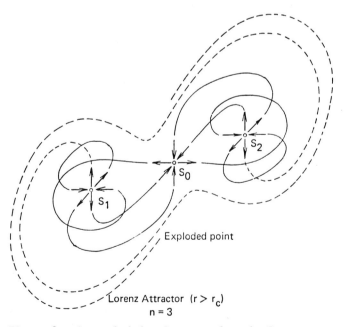

Figure 2.   An exploded point example, the Lorenz attractor.

		S_0		S_1	S_2	LA
		s	s	s	s	s
S_0	u	1	1	(1)	(1)	1
	u	1	(1)	(1)	1	1
S_1	u	(1)	1	1	1	(1)
	u	1	(1)	1	(1)	1
S_2	u	(1)	1	1	1	(1)

Figure 3. Origin and destination of the half-flows.

Here **n** is the number of columns in Figure 3, $K_n = K_{n-2}+3$, for $n \geq 5$ and $K_1 = 0, K_2 = 1, K_3 = 2, K_4 = 3.$

Example 2. Rössler Attractor (Independent Peeling) [5]

$$dx/dt = -y-z$$

$$dy/dt = x+ay$$

$$dz/dt = b+xz-cz$$

$$CLS(x) = (x_1, \; 3w_{m1})$$

A new singular point $x_{2,3}$ is created as a second generating point. Independent peeling of x_1 and $x_{2,3}$ separately yields

$$CLS'(x) = (x_1, \; 2w_{m1} + 1\,w_{m2}, \; E_1, \text{ exploded point})$$

$$(x_{2,3}, \text{ double point } 1w_{m1} + 2\,w_{m2}$$

$$E_2, \text{ exploded point if } x_{2,3} \text{ elliptic}$$

$$\text{or } L_3, \text{ limit cycle if } x_{2,3} \text{ hyperbolic})$$

Subsequent bifurcation results in the critical set,

$$CLS''(x) = (x_1, E_1, \; x_{2,3}, E_2 \text{ (or } (L_2)\text{, } L_3))$$

Globally $x_1 = 2w_{m1} + 1w_{m2}$

$$x_2 = 1w_{m1} + 2w_{m2}$$

$$x_3 = \underline{1w_{m1} + 2w_{m2}}$$

$$4w_{m1} + w_{m2}$$

which implies an exploded point E. However, it is decomposed into the elements found above, i.e.

$$E = E_1 + E_2 + L_3$$

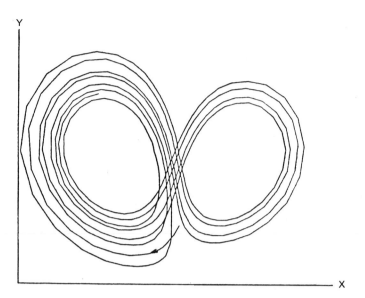

Figure 4.   An exploded point in the Rössler attractor.

and possibly there appears an unstable limit cycle L_3^u balancing the stable
one L_3.   One of the exploded points is shown in Figure 4.

Example 3.   Toroidal Attractor   (Peeling to Noncritical Set), [8]

$$dx/dt = x - ay - xz$$

$$dy/dt = bx + y - yz$$

$$dz/dt = x^2 + y^2 + cz - (z/(z+d))$$

Bifurcation of the generating singular point first to a CLS(x) and then to
a NCLS(x), is depicted in Figure 5.   In this noncritical set, bifurcation
of the second order limit cycle takes place to yield another limit cycle
which forms a toroidal surface, Figure 6.

Example 4.   Combined Peeling [6]

$$dx_1/dt = x_2$$

$$dx_2/dt = -p_2x_1 - p_1x_2 - x_1^3 - x_1^2x_2$$

It should be emphasized that, in addition to the above variations in the
bifurcation to oscillating or nonoscillating solutions, in the multi-
parameter case peeling as one parameter varies might be independent from

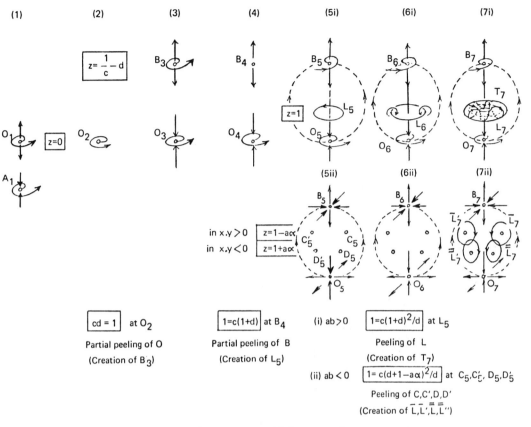

Figure 5: Bifurcations leading to toroidal attractor.

that taking place as another parameter varies. This is illustrated by the present example. In this case there are two bifurcation parameters p_1 and p_2. The system starts with a stable S_0 for p_1, $p_2 > 0$. For $p_1 > 0$, as p_2 varies, peeling into three singular points takes place for $p_2 < 0$.

$p_1 > 0$, $p_2 < 0$:

S_0 becomes a saddle, S_1 and S_2 (stable foci) are created.

$p_1 < 0$, $p_2 > 0$:

While $p_2 > 0$, as p_1 becomes negative the singular point peels to yield an unstable singular point and a surrounding stable limit cycle.

S_0 becomes an unstable focus and a stable LC is created.

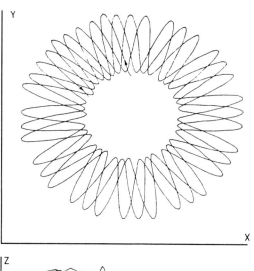

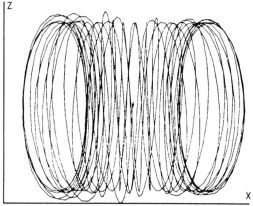

Figure 6.   Toroidal attractor.

$p_1 < 0, \; p_2 < 0$:

When both parameters exceed the respective bifurcation values, $p_1 = 0$ and $p_2 = 0$, the two separate bifurcation results are combined to yield multiple singular solutions, as well as the stable limit cycle. Moreover, there appear various possibilities of additional limit cycles being created, due to the stability changes at the bifurcating solutions S_1 and S_2.

A detailed bifurcation study of the system may easily be carried out, resulting in further information. In detecting the additional limit cycles, particularly the unstable ones, the global stability theorem above and the concept of index (and f-dimension) may be applied. Therefore in the region $p_1 < 0$ and $p_2 < 0$, as the solutions S_1 and S_2 bifurcate, the index of an expected oscillatory solution can be calculated. It is further observed that this analysis shows also that S_1 and S_2 peel (cooperatively) just as in the case of the Lorenz example. However, the significant aspect of the example is to illustrate the combined peeling phenomenon, which is not observed in the case of either the Lorenz or the other examples discussed above.

V. CONCLUSIONS [9,10]

We discussed the following classes of peeling:

. cooperative peeling
. independent peeling
. peeling to a noncritical limiting set
. combined (complex) peeling

Also we demonstrated that the stability index defined in this paper is useful in the analysis of oscillating solutions of a dynamical system.

We believe that, although indices related to dimensions based on stochastic considerations (e.g. see [9]) are informative, they may not be as fundamental as the stability index discussed herein, which is based on a "deterministic" dimension. Such a deterministic index reveals more basic information in classifying characteristic solutions than stochastic indices.

In Neimark's work (e.g. see [10]), we find an approach which appears to be similar to the notion of the f-dimension on which our stability index is based.

ADDENDUM

An extensive list of oscillatory solutions with rich behavior in fields related to chemical systems is given in Reference [11]. A classification of these models may be based on the present discussion. In addition, in a recent publication [12], a classification has been discussed and related to applications in celestial mechanics.

REFERENCES

I. A STABILITY INDEX

1.   O. Gurel, A Classification of Singularities of (X,f) Mathematical
        Systems Theory, vol.7,no.2(1973)154-163.
2.   _____, Exploded Points, Z. Naturforschung 36a (1981) 72-75.
3.   _____, Decomposed Partial Peeling and Limit Bundles, Physics
        Letters, 61A, n.4. (1977)219-223.

II. COOPERATIVE PEELING

4.   _____, Necessary and Sufficient Conditions for Cooperative Peeling
        of Multiple Generating Singular Points, In: Bifurcation
        Theory and Applications in Scientific Disciplines, The New
        York Academy of Sciences Annals No. 316, (O. Gurel & O. E.
        Rössler, eds.)(1979).
5.   _____, Individual Peeling of Multiple Singular Points,
        Naturforschung 36a (1981) 311-316.

III. OTHER CLASSES OF PEELING

6.   _____, On the Cardinality, Stability and Dimensionality of
        Oscillating Solutions, In: Systems Science and Science(Bela
        H. Benathy, ed.) Society for General Systems Research,
        (1980)266-270.

IV. EXAMPLES

7.   _____, Poincaré's Bifurcation Analysis, In: Bifurcation Theory and
        Applications in Scientific Disciplines, pp. 5-26.
8.   _____, & O. E. Rössler, Bifurcation to Toroidal Surfaces,
        Mathematica Japonica, vol.23, n.5, (1979) 491-507.
9.   D. A. Russel, J.D. Hanson & E. Ott, Dimension of Strange Attractors,
        Physical Review Letters, vol. 45, n. 14 (1980) 1175-1178.
10. Yu. I. Neimark, Invarianthie Mnogoobraziya i Stokhasticheskie
        Dvijeniya Dinamicheskikh Sistem, (Invariant Manifolds and
        Stochastic Motion of Dynamical Systems),
        In: Problemi Asimptoticheskoy Teorii Nelineynikh Kolebaniy
        (Problems of the Asymptotic Theory of Nonlinear
        Oscillations), Kiev, Naukova Dumka, 1977.

REFERENCES ADDED IN PROOF:

11. O. Gurel and Demet Gurel, Types of Oscillations in Chemical
        Reactions, In: Topics in Current Chemistry, Springer-Verlag
        Vol. 118 (1983) pp. 1-73.
12. _____, Exploding Dynamical Systems, In: Applications of Modern
        Dynamics to Celestial Mechanics and Astrophysics (V.
        Szebehely, Ed.) D. Reidel Publ. Co., (1982) 277-299.

5

Unfolding of Degenerate Bifurcations

W. F. Langford *

Department of Mathematics
McGill University
Montreal, Quebec, Canada

Two classes of degenerate bifurcations are described, together with their unfoldings. The universal unfoldings of degenerate Hopf bifurcations can be completely analysed using singularity theory methods. The unfoldings of coalesced simple and Hopf bifurcations can give rise to secondary bifurcations including the bifurcation of a torus and chaotic dynamics. A numerical example is presented.

1. INTRODUCTION

In the long history of the development of bifurcation theory, dating from the pioneering work of Euler [7] and Poincaré [24], it is only recently that a systematic study of degenerate bifurcations has been undertaken. Important motivation for this study was provided by Bauer, Keller and Reiss [1], who showed that multiple eigenvalue bifurcations (a degeneracy often encountered in applications) can lead to secondary bifurcation (an elusive phenomenon of considerable practical interest). Their analysis

*Present address: Department of Mathematics and Statistics, University of Guelph, Guelph, Ontario, Canada

was refined by Keener [16], whose results lead to the investigations of multiple bifurcations described here, in section 4 and 5. Another type of degeneracy occurs when eigenvalues are simple, but generic nonlinear terms are missing.   Takens [27], and Chafee [4], investigated the Hopf bifurcation with this type of degeneracy. A recent extension of their work is reviewed in section 3.

These investigations are parallel to, though they proceeded independently of, the development of catastrophe theory. Recently, for the case of steady-state bifurcations, the link with catastrophe theory via fundamental methods of singularity theory has been securely welded by Golubitsky and Schaeffer [9]. Smale [26] has observed that the Hopf bifurcation lies "deeper" than catastrophe theory because of such essentially dynamical aspects as the exchange of stabilities; yet even here singularity theory methods have yielded new information, see [8] and section 3.

After summarizing two historically important nondegenerate bifurcations in section 2, and reviewing some recent work on degenerate bifurcations in 3 and 4, this paper presents in section 5 new numerical studies of a multile bifurcation arising from the coalescence of a Hopf bifurcation with a simple bifurcation of steady-states. The computer-generated graphics clearly illustrate the phenomena of secondary Hopf bifurcation and of the bifurcation of an invariant torus of solutions, the existence of which was first demonstrated in [18]. The significance of this example is that it is obtained by truncation of normal form equations which can occur generically (under the specified conditions) in dynamical systems of dimension 3 or more, even infinite-dimensional. Therefore it is reasonable to anticipate that the behavior displayed by this example will be found in real applied problems, for example in reaction-diffusion equations or fluid mechanics [2, 10, 15, 16].

2. NONDEGENERATE BIFURCATIONS

A nondegenerate bifurcation is, roughly speaking, a bifurcation which is preserved qualitatively under all small admissible perturbations of the problem.   See [20] for a more precise definition. In this section we consider two important nondegenerate bifurcations: the so-called simple bifurcation and the Hopf bifurcation.

The history of simple bifurcation begins at least as early as Euler's study of the buckling of rods [7]. The following is a fairly recent theorem due to Crandall and Rabinowitz [5]. Consider an equation

$$F(u,\lambda) = 0 \qquad\qquad F \ \varepsilon \ C^2 \colon X \times \mathbb{R} \to Y \tag{1}$$

where X and Y are Banach spaces and (1) has the trivial solution

$$F(0,\lambda) = 0 \qquad\qquad \lambda \ \varepsilon \ \mathbb{R} \tag{2}$$

In this paper, only F satisfying (2) are permitted. The "imperfect" case, in which perturbations violating (2) are allowed, is also well understood; see [2, 3, 9, 17]. Let A denote the Fréchet derivative $D_u F$ at $(u,\lambda) = (0,\lambda^o)$, with nullspace and range N(A) and R(A) respectively.

Theorem (Simple bifurcation):

If (a) N(A) is one-dimensional, spanned by ϕ

  (b) Codim R(A) = 1   ($\equiv$ dim(Y/R(A)))

  (c) $D_{u\lambda} F(0,\lambda^o)\phi \ \varepsilon \ R(A)$

then there exist nontrivial solutions of (1) near $(0,\lambda^o)$, parameterized by x for small $|x|$, of the form:

$$u(x) = x\phi + xw(x) \qquad\qquad w(0) = 0 \tag{3}$$
$$\lambda(x) = \lambda^o + xv(x)$$

where w and v are continuous for small $|x|$, v is real and w lies in the complement of the span of ϕ. The branch (3) is unique up to parameterization.

The proof of this theorem proceeds by the "Liapunov–Schmidt" method which splits the equations $F(u,\lambda) = 0$ into an invertible part in R(A) to which the implicit function theorem applies, and a one-dimensional singular part, known as the bifurcation equation, which yields the unique nontrivial branch of solutions.

The Hopf bifurcation theorem gives conditions for the creation of a limit cycle from an equilibrium point of a differential equation as a parameter varies. Consider for example an ordinary differential equation in n-space, parametrized by λ:

$$F(u,\lambda) = du/dt - f(u,\lambda) = 0 \tag{4}$$

where $f \ \varepsilon \ C^3$ and we assume that an equilibrium solution has been translated to the origin, i.e. $f(0,\lambda) = 0$, $\lambda \ \varepsilon \mathbb{R}$.

Theorem (Hopf bifurcation):

If (a) $D_u f(0,\lambda)$ has simple complex eigenvalues a ± ib satisfying, at λ = λ^0, a = 0 and b ≠ 0 and no other eigenvalues are on the imaginary axis

(b) $a'(\lambda^0)$ ≠ 0 (the strict crossing condition)

then there exists a branch of periodic solutions of (4), parameterized by x for small $|x|$, of the form:

$$u(s,x) = x\phi(s) + x^2 w(s,x)$$
$$\lambda(x) = \lambda^0 + x^2 v(x) \tag{5}$$
$$T(x) = 2\pi/b + x^2 P(x)$$

where T is the period, s = $2\pi t/T(x)$, $\phi(s)$ = Re[Vexp(is)], V is the eigenvector corresponding to ib, w lies in the complement of span[ϕ,ϕ'] and is continuous in x, and v and P are continuous even functions of x. The solution branch (5) is unique up to phase shift.

Many proofs of this theorem are known, using various methods such as Liapunov-Schmidt [6], center manifold [23], averaging [13], Birkhoff normal forms [14], etc.

3. DEGENERATE HOPF BIFURCATION

The primary question of practical interest where Hopf bifurcation is concerned is the "direction" of the bifurcation, that is whether the limit cycle occurs for λ greater than or less than λ^0. Locally, this question is answered by the sign of v(0) in (5), except in the degenerate case v(0) = 0. Then the computation of v(0) (by well-known formulae) fails to answer this question, and higher-order terms in $\lambda(x)$ have been computed for this purpose (see [8] and further references therein). A second degeneracy which may arise in the Hopf bifurcation theorem is the violation of condition (b), i.e. $a'(\lambda^0)$ = 0, and here again various authors have investigated the implications of vanishing higher order derivatives $a'(\lambda^0)$ and have obtained generalizations of the Hopf theorem.

However, it should be remarked that generic perturbations of (4) will remove either of these two types of degeneracy, and such perturbations are virtually unavoidable in applications, so that the perturbations of these degenerate Hopf bifurcations are of greater practical importance than the

TABLE 1

Defining Conditions	Nondegeneracy Conditions	Normal Form Universal Unfolding	Codimension
none	$G_\xi \neq 0 \neq G_\lambda$	$x^3 \pm \lambda x = 0$	0
$G_\xi = 0$	$G_{\xi\xi} \neq 0 \neq G_\lambda$	$x^5 + 2hx^3 \pm \lambda x = 0$	1
$G_\lambda = 0$	$G_\xi \neq 0 \neq G_{\lambda\lambda}$	$x^3 \pm (\lambda^2 + h)x = 0$	1
$G_\xi = 0 = G_{\xi\xi}$	$G_{\xi\xi\xi} \neq 0 \neq G_\lambda$	$x^7 + kx^5 + hx^3 \pm \lambda x = 0$	2
$G_\lambda = 0 = G_{\lambda\lambda}$	$G_\xi \neq 0 \neq G_{\lambda\lambda\lambda}$	$x^3 \pm (\lambda^3 + k\lambda + h)x = 0$	2
$G_\xi = 0 = G_\lambda$	$G_{\xi\xi} \neq 0 \neq G_{\lambda\lambda}$ and $d \neq 0, \pm 1$	$x^5 + 2d\lambda x^3 +$ $[\lambda^2 + sgn(d)k\lambda + h]x = 0$	3
$G_\xi = 0 = G_\lambda$	$G_{\xi\xi} \neq 0 \neq G_{\lambda\lambda}$ and $d \neq 0$	$x^5 + 2d\lambda x^3 -$ $[\lambda^2 + sgn(d)k\lambda + h]x = 0$	3

degenerate bifurcations themselves. This is the point of view adopted in [8], where the following new results were obtained.

Assuming $f \in C^\infty$, we can write

$$\lambda(x) = \lambda^0 + \sum_{j=1}^{n+1} x^{2j} v^{(2j)} + o(x^{2n+2}) \tag{6}$$

Then consider the degenerate cases:

$$v^{(2)} = v^{(4)} = \ldots = v^{(2n)} = 0, \; v^{(2n+2)} \neq 0$$
$$a^{(j)}(\lambda^0) = 0, \; j = 1, \ldots, m; \; a^{(m+1)}(\lambda^0) \neq 0 \tag{7}$$

while retaining hypothesis (a) in the Hopf theorem. Using classical methods it is straightforward, although laborious, to compute the resulting bifurcations. Using singularity theory it is possible to compute in addition the "universal unfoldings" which give all possible perturbations of these degenerate bifurcations. The procedure, as used in [8], is to reduce (4) via the Liapunov-Schmidt method, and take advantage of the SO(2) symmetry, to obtain a single scalar equation of the form

$$g(x,\lambda) = xG(x^2,\lambda) = 0 \qquad\qquad (8)$$

where g,G are germs of C^{∞} functions mapping ($\mathbb{R}^2$,0) $\rightarrow \mathbb{R}$, G(0,0) = 0, and g is odd in x. For convenience, write $\xi = x^2$. Then the conditions (7) translate directly into corresponding conditions on G, which together with conditions on modal parameters lead to a classification of g and its unfoldings, up to symmetry-covariant contact equivalence, represented by polynomial normal forms; see Table 1 for some examples of codimension $\leq$ 3. The corresponding bifurcation diagrams can be found in [8]. Here $d = G_{\xi\lambda}/|G_{\xi\xi} \cdot G_{\lambda\lambda}|^{1/2}$ is a modal parameter, and h, k are unfolding parameters.

4. COALESCENCE OF SIMPLE AND HOPF BIFURCATIONS

It is not unusual in applications for the same differential equation to have both a simple bifurcation of steady-states and a Hopf bifurcation of a limit cycle, for different values of the bifurcation parameter λ. Then, by varying a second parameter σ(the "splitting" parameter), these two primary bifurcations can be made to coalesce at $\lambda = \lambda^o$. See Figure 1, where z and r are amplitudes as in (11) below.

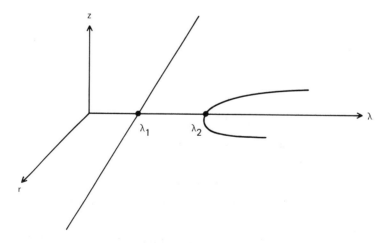

Figure 1.   Simple bifurcation at $\lambda = \lambda_1$.
             Hopf bifurcation at $\lambda = \lambda_2$.

More precisely, let us consider a differential equation

$$du/dt = f(u,\lambda,\sigma) \qquad\qquad f: \ X \times \mathbb{R}^2 \to Y \supset X \qquad\qquad (9)$$

where we assume $f \in C^2$, $f(0,\lambda,\sigma) = 0$, and $D_u f$ is Fredholm with simple eigenvalues $a \pm ib$ and c (a,b,c real), at $(u,\lambda,\sigma) = (0,0,0)$, satisfying

$$a = 0 = c, \ b \neq 0, \quad \frac{\partial(c,a)}{\partial(\lambda,\sigma)} = \begin{vmatrix} d_1 & d_2 \\ d_3 & d_4 \end{vmatrix} \neq 0 \qquad\qquad (10)$$

With additional nondegeneracy conditions on the second derivatives of f at (0,0,0), as specified in [18], we obtain the normal form bifurcation equations

$$\begin{aligned}
dz/dt &= (d_1\lambda + d_2\sigma)z + pz^2 + qr^2 + \cdots \\
dr/dt &= (d_3\lambda + d_4\sigma)r + mrz + \cdots \\
d\theta/dt &= b + \cdots
\end{aligned} \qquad\qquad (11)$$

Here z and r are amplitudes of projections on the eigenspaces of the eigenvalues 0 and ib associated with the simple and Hopf bifurcations respectively, θ is the phase, an p, q, m are computable constants. The solutions of (11) are easily analysed, and locally determine the solutions of the original equations (9). There are six distinct bifurcation diagrams arising from (11), every one of which includes a secondary Hopf bifurcation, see [18]. The most interesting of the six cases, shown in Figure 2, has an additional bifurcation, that of an invariant torus of solutions from the limit cycle. This torus bifurcation can be thought of as coming from a Hopf bifurcation in the (z,r) amplitude equations (11). Solutions on the torus have two asymptotic frequencies, the first is near b from the original Hopf bifurcation, and the second is small (order σ) from the Hopf bifurcation in (11). The torus bifurcation for (11) degenerates when one omits the higher order terms indicated by dots, because then the equations (11) are exactly integrable, see [2, 10, 11, 15, 17]. The result in this case is an infinite family of nested tori, all existing at the same value of the bifurcation parameter λ. However generic cubic terms remove this degeneracy and give a torus which grows in diameter as the bifurcation parameter varies. (This torus is studied numerically in the next section). The torus is confined initially by the unstable manifold of the upper equilibrium point, as long as this manifold

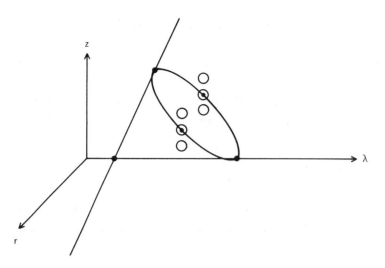

Figure 2.   Interactions of simple and Hopf bifurcations.
             Circles represent torus bifurcation.

lies inside the basin of attraction of the lower equilibrium point. If we
neglect nonaxisymmetric terms in (11), then as the torus grows in diameter
it approaches coalescence with a saddle connection manifold (separatrix)
joining the two saddle points. In the limit, the period of the secondary
motion around the torus is infinite. Beyond this point, the torus bursts
and the orbits escape to negative infinity along the z-axis. (This is
closely analogous to Abraham's "blue sky" catastrophe).

Of course this limiting coalescence is peculiar to the axisymmetric
case. In the generic nonaxisymmetric case, transversal intersection of
these stable and unstable manifolds of the two saddle points will occur,
leading to heteroclinic orbits, Smale horseshoes and hence to "chaos", see
[10, 20, 21].

A similar torus bifurcation can occur when a Hopf bifurcation is near
coalescence with steady-state bifurcations of pitchfork, saddlenode, or
hysteresis types; for details see [11, 15, 20, 22].

5. NUMERICAL EXAMPLE

We now present the results of some computer studies of a model example of
degenerate bifurcation, satisfying the conditions for (11). This model is
derived from an example in [18] which in turn was based on a model for
turbulence due to Hopf. It is interesting to note that Hopf's model was
designed to illustrate the so-called Landau-Hopf route to turbulence in

which there is an infinite succession of Hopf-type bifurcations yielding
quasiperiodic flows with successively higher numbers of frequencies, and
thus a gradual transition to turbulence. However, today we know that
turbulence-like behavior can arise after only three such bifurcations, as
for example in the Ruelle-Takens route to turbulence [25], and in fact the
mechanism of the previous section and the present example leads to chaotic
(weakly turbulent) behaviour after the appearance of only two fundamental
frequencies.

Consider the differential equations with $(x,y,z) \in R^3$

$$dx/dt = (\lambda - \sigma)x - by + x[z + c(1 - z^2)]$$
$$dy/dt = bx + (\lambda - \sigma)y + y[z + c(1 - z^2)] \qquad (12)$$
$$dz/dt = \lambda z - (x^2 + y^2 + z^2)$$

This example satisfies the conditions of the previous section leading to
the normal form (11) and the bifurcation diagram in Figure 2. The
degeneracy of the torus bifurcation referred to above is removed by the
cubic terms for $c \neq 0$. However, note that (12) is still symmetric about
the z-axis. For the effects of nonaxisymmetric perturbations of (12), and
further study of the resulting chaotic behavior, see [21].

Computed phase portraits of (12) are shown in Figures 3 to 10. The
bifurcation analysis of the previous section is valid only locally, that
is for small (x,y,z,λ,σ). For convenience we have blown up these
variables to order 1 by rescaling, the effect on the equations (12) is to
make the cubic coefficient c small, the frequency b larger and to make t a

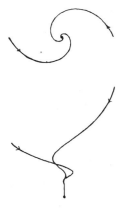

Figure 3. Stable trivial steady-state solution, $\lambda = -0.2$.
           t-intervals [0.50] and [50,100].

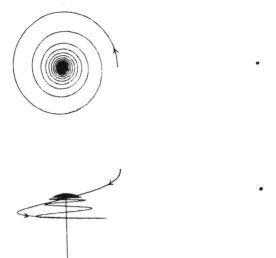

Figure 4.   Stable nontrivial steady-state solution, $\lambda = 1.6$.
            t-intervals $[0,50]$ and $[50,100]$.

slow time variable. In these figures, $\sigma = 3.0$, $c = 0.2$, $b = 4.0$ (except
in Figures 6 and 8) and λ increases in the range from approximately 0 to
2. The left and right plots of each pair show the transient behavior on
an initial t-interval and the convergence to the attractor on a later t-
interval, for the same solution in each case. The upper and lower pairs
show the (x,y) and (x,z) projections respectively of the same solution of
(12). The origin is marked by a dot in each plot.

For $\lambda < 0.0$, the origin is stable as shown in Figure 3. (Stable for
this discussion means asymptotically stable or orbitally asymptotically
stable as appropriate). A bifurcation of steady-states occurs at $\lambda = 0.0$,
with an exchange of stabilities, so that for small $\lambda > 0.0$ the origin is
unstable and the steady-state with $z > 0.0$ is stable, see Figure 4. This
latter steady-state undergoes a (secondary) Hopf bifurcation at
approximately $\lambda = 1.68$, yielding a stable limit cycle as shown in Figure
5. At $\lambda = 2.0$ this limit cycle becomes a "vague attractor", with Floquet
multipliers on the unit circle so that it has neutral linearized stabiity,
and only the higher order terms render it attracting. Figure 6 shows the

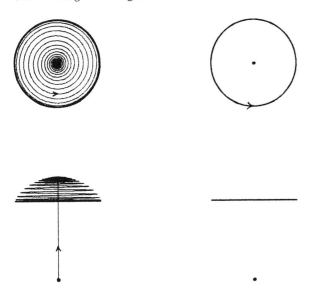

Figure 5.   Stable limit cycle, $\lambda = 1.8$.
            t-intervals [0,50] and [50,100].

very slow approach to the cycle in this case. Here, as in Figure 8, we
have $b = 0.25$ in order to observe more clearly the approach to the
attractor, by permitting observation over a longer time interval. Even
so, the decay of transients is far from complete in Figure 6. For $\lambda >$
2.0, an attracting invariant torus appears, which grows rapidly in
diameter until it "bursts", see Figures 7 to 9, and 10.

Figures 11 to 13 are plots of the fast Fourier transforms of the same
solutions depicted in Figures 5, 7 and 9 respectively, after the decay of
transients. They show the transition from periodic to quasiperiodic
behavior. Note that all the frequencies present in Figures 12 and 13 are
integer combinations of two fundamental frequencies.

A remarkably similar example of a singular point of a vector field
exhibiting multiple bifurcations, including the bifurcation of a toroidal
surface of solutions captured inside a two dimensional separatrix surface,
has been studied indepdendently by Gurel and Rössler [12]. However, their
hypotheses on the vector field exclude the occurrence of a Hopf
bifurcation from the trivial solution $(x,y,z) = (0,0,0)$, which puts their
example outside the scope of the analysis of this paper.

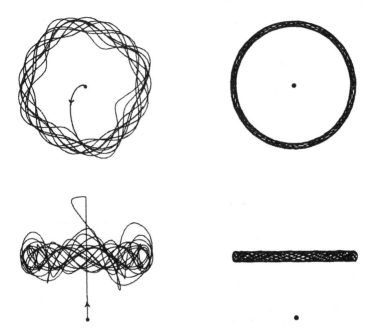

Figure 6.   Vague attractor, λ = 2.0, **b** = 0.25 .
            t-intervals [0,250] and [750,1000].

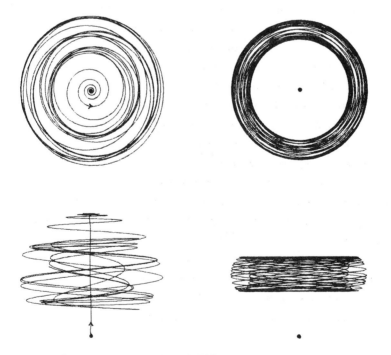

Figure 7.   Thin torus, λ = 2.003 .
            t-intervals [0,40] and [640,680].

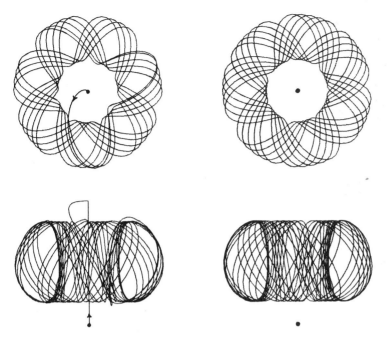

Figure 8.   Torus, λ = 2.015, **b** = 0.25 .
            t-intervals [0,250] and [250,500].

Figure 9.   Torus near separatrix and infinite period bifurcation,
            λ = 2.0248, **t**-intervals [0,100] and [200,300].

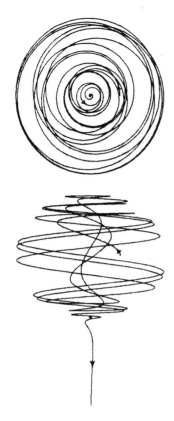

Figure 10. Exploded torus after infinite period bifurcation,
$\lambda = 2.0249$, $t \epsilon [0,100)$.

Figure 11. Fourier transform of solution in Figure 5.

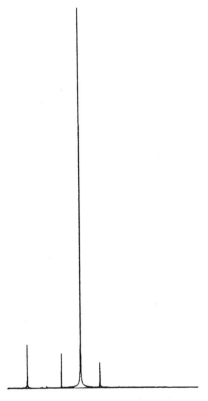

Figure 12. Fourier transform of solution in Figure 7.

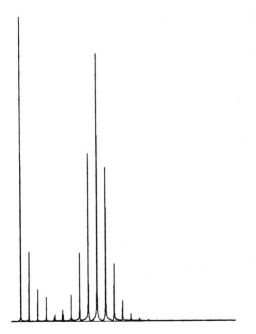

Figure 13. Fourier transform of solution in Figure 9.

REFERENCES

1. L. Bauer, H. B. Keller and E.L. Reiss. Multiple eigenvalues lead to
    secondary bifurcation. SIAM Rev. 17 (1975) 101-122.
2. E. J. Bisset. Interaction of steady and periodic bifurcating modes
    with imperfection effects in reaction–diffusion systems. SIAM J.
    Appl. Math. 40 (1981) 224-241.
3. J. -F. Boivin. Catastrophe theory and bifurcations. M.Sc. thesis,
    Department of Mathematics, McGill University (1981).
4. N. Chafee. Generalized Hopf bifurcation and perturbation in a full
    neighborhood of a given vector field. Indiana Univ. Math. J. 27
    (1978) 173-194.
5. M. G. Crandall and P. H. Rabinowitz. Bifurcation from simple
    eigenvalues. J. Funct. Anal. 8 (1971) 321-340.
6. M. G. Crandall and P. H. Rabinowitz. The Hopf bifurcation theorem.
    MRC Tech. Summary Report No. 1604, Mathematics Research Center,
    University of Wisconsin, Madison (1976).
7. L. Euler. Methodus inveniendi lineas curvas maximi minimi minimivi
    proprietate grudentes, Lausanne (1744). In Opera Omnia I, V. 24,
    231-297. Fussli, Zurich (1960).
8. M. Golubitsky and W. F. Langford. Classification and unfoldings of
    degenerate Hopf bifurcations. J. Diff. Eq. 41 (1981) 375-415.
9. M. Golubitsky and D. Schaeffer. A theory for imperfect bifurcation
    via singularity theory. Commun. Pure Appl. Math. 32 (1979) 21-98.
10. J. Guckenheimer. On a codimension two bifurcation. In Dynamical
    Systems and Turbulence, Lecture Notes in Mathematics No. 898,
    Springer-Verlag (1981) 99-142.
11. J. Guckenheimer. Multiple bifurcation problems of codimension two.
    University of California, Santa Cruz. Preprint (1981).
12. O. Gurel and O.E. Rössler. Bifurcation to toroidal surfaces. Math.
    Japonica 23 (1979) 491-507.
13. J. K. Hale. Generic bifurcation with applications. In Nonlinear
    analysis and mechanics: Heriot-Watt Symposium Vol. I. Pitman,
    London (1977) 59-157.
14. G. Iooss and D.D. Joseph. Elementary Stability and Bifurcation
    Theory. Springer-Verlag, New York (1980).
15. G. Iooss and W.F. Langford. Conjectures on the routes to turbulence
    via bifurcations. In Nonlinear Dynamics, R.H.G. Helleman editor.
    Ann. N.Y. Acad. Sci. 357 (1980) 489-505.
16. J. P. Keener. Secondary bifurcation in nonlinear diffusion reaction
    equations. Studies in Appl. Math. 55 (1976) 187-211.
17. J. P. Keener. Infinite period bifurcation and global bifurcation
    branches. SIAM J. Appl. Math. 41 (1981) 127-144.
18. W. F. Langford. Periodic and steady-state mode interactions lead to
    tori. SIAM J. Appl. Math. 37 (1979) 22-48.
19. W. F. Langford. Modulation instability and weak turbulence.
    Dynamical Systems II, A.R. Bednarek and L. Cesari editors. Academic
    Press, N. Y. (1982) 589-593.
20. W. F. Langford. A Review of interactions of Hopf and steady-state
    bifurcations. To appear in nonlinear Dynamics and Turbulence,
    G. Iooss and D.D. Joseph editors. Pitman Press.
21. W. F. Langford. Chaotic dynamics in the unfoldings of degenerate
    bifurcations. Proceedings, International Symposium on Applied
    Mathematics and Information Science, Kyoto University, Kyoto,
    Japan, March 29-31, 1982.

22. W. F. Langford and G. Iooss. Interactions of Hopf and pitchfork
    bifurcations. In Bifurcaton Problems and their Numerical Solution,
    H.D. Mittelmann and H. Weber editors. ISNM 54, Birkhauser-Verlag,
    Basel (1980) 103-134.
23. J. E. Marsden and M. McCracken. The Hopf Bifurcation and its
    Applications. Springer-Verlag, New York (1976).
24. H. Poincaré. Sur l´equilibre d´une masse fluide animée d´un mouvement
    de rotation. Acta Mathematica 7 (1885) 259-380.
25. D. Ruelle and F. Takens. On the nature of turbulence. Comm. Math.
    Phys. 20 (1970) 167-192.
26. S. Smale. Book Review of: Catastrophe Theory, Selected Papers
    (E.C. Zeeman). Bull. Amer. Math. Soc. 84 (1978) 1360-1368.
27. F. Takens. Unfoldings of certain singularities of vector fields:
    generalized Hopf bifurcations. J. Diff. Eq. 14 (1973) 476-493.

6

Example of an Axiom A ODE

Otto E. Rössler

Institute for Physical and Theoretical Chemistry
University of Tübingen
Tübingen, Federal Republic of Germany

A 4-variable ordinary differential equation is described which possesses an axiom A attractor in the sense of Smale. The assumptions are that some of its parameters are sufficiently small, and that a certain map-theoretic conjecture applies. The cross-section is a generalized solenoid. The example presented shows that strange attractors can be expected in realistic systems.

1. INTRODUCTION

In 1967, Smale described a class of mathematically simple systems producing non-trivial ("chaotic" is the current term) behavior in a structurally stable manner. An explicit example was the solenoid diffeomorphism (see also Smale, 1977, for an explicit difference equation) which produces an axiom A attractor. This map is the simplest analogue to Hopf's (1937) baker's transformation that meets the additional constraints of being both contracting and differentiable (Rössler, 1980).

Contracting axiom A diffeomorphisms are mathematical formalizations of a taffy-puller, that is, a mechanical machine that mixes taffy (caramel mass) by successively stretching and folding it. Hereby, adjacent points diverge exponentially while non-adjacent ones may come close together ('mixing'). The solenoid is obtained by first taking a ring (solid torus) of taffy, then elongating it to twice its length while at the same time shrinking its volume (due to squeezed-out fat, for example), then wrapping it up once so that it becomes double-looped, and finally putting it back into the inside of its original location. The trick with this volume-contracting differentiable map is that it is hyperbolic everywhere in the sense of showing contraction in one direction and elongation in another. This uniformity makes the mathematics simpler (and indeed allows the formulation of specific sub-axioms that together comprise axiom A; see Smale, 1967, for the details of the abstract definition).

Surprisingly, two expectations that were attached to the original formalism did not bear out. The first was that an axiom A attractor could not possibly be generated in a two-dimensional map (and hence in a three-dimensional flow). Plykin (1974) described a recipe to generate a two-dimensional analogue of the solenoid diffeomorphism. His non-explicit map contains 9 fixed points (6 saddles, 3 unstable nodes) in a rather rigidly prescribed pattern. This means that the odds for ever finding an explicit equation for such a map (or an O.D.E. governed by such a map, respectively) are extremely low. In other words, axiom A attractors cannot be expected to be found in O.D.E.'s of three dimensions.

The second implicit hypothesis was that axiom A flows might nonetheless be frequent. So far, the not – everywhere – differentiable competitor to axiom A – the class to which the contracting baker's transformation belongs – has been the only class for which explicit O.D.E.'s have been found. All of these equations either produce a Lorenz attractor or an analogue of it (cf. Rössler and Ortoleva, 1978). The Lorenz attractor (Guckenheimer, 1976; similarlly Rössler, 1976) shares a number of features with an axiom A attractor, but is not generated by a diffeomorphism. Another difference is its lack of structural stability (Williams, 1979).

In the following, a 4-variable O.D.E. producing an axiom A attractor is proposed.

2. A CLOCKS GAME

Two old watches (that may be broken and deprived of one hand, but should still be resettable) can be employed in a 'resetting game.' One

person plays the ´metronome´, saying "right, left, right, left, ..." with
appropriate pauses in between. The person holding the ´right´ watch
resets the latter to twice the time shown by the left watch when being
addressed. Vice versa for the third player who holds the ´left´ watch.
This three-person game comes close to implementing an axiom A flow – if
only one could show that persons are differentiable.

The metronome player can in principle be replaced by an autonomous
oscillator, and for each of the two watch-carrying persons, a two-variable
nonlinear O.D.E. may be substituted. The resulting system is still a
candidate for an axiom A system – albeit now in the form of a 6-variable
autonomous O.D.E.

Even though such a deterministic system has too many variables to be
readily understood, it can be used as a conceptual model with which to
check a number of preliminary questions. For example, one knows that no
´resetting´ can ever be made absolutely ´to the point´ – there must be a
remaining smooth deviation. Moreover, the resetting functions to be used
(the trigonometric formulas for computing the rectangular coordinates of
an angle 2α from those of α; namely, $y´ = 2xy$, $x´ = y^2 - x^2$ since $\sin 2\alpha = 2 \sin\alpha \cos\alpha$ and $\cos 2\alpha = \sin^2\alpha - \cos^2\alpha$) will in general be realizable only
approximately – which is tantamount to assuming the presence of additional
perturbing terms of low order. It makes sense to ask whether either of
these assumptions (or their combination) destroys the game by producing an
attracting periodic solution. It appears plausible that the strong
stretching (by a factor of two) that in each half-map applies in a lateral
direction will ´override´ all other (shearing and compressing) modulating
effects as long as these are minor. This conjecture can be tested by
writing down the next-angle function for the two clocks explicitly:

$$\alpha_{n+2} = 2\beta_n + \varepsilon \sin(\alpha_n - 2\beta_n) + \delta \ f_1(\alpha_n, \beta_n)$$

$$\beta_{n+2} = 2\alpha_{n+2} + \varepsilon \sin(\beta_n - 2\alpha_{n+2}) + \delta \ f_2(\beta_n, \alpha_{n+2}),$$

$$(1)$$

where the first perturbing term represents an idealized (sinusoidal)
deviation law. Note that the term α_{n+2} in the second line in each case
stands for all the terms of the right hand side of the first line (as a
convenient shorthand). There are no n+1 terms present. A relabeling of
n+2 terms toward n+1 is possible, but has been avoided in order to make
the straightforward derivation of Eq.(1) easier to check. The two major

questions are (1) whether the unique steady state (a saddle) persists as such, and (2) whether the property of lateral expansion also persists (presumably in an oblique, 'precessing' manner) if both ε and δ are given small enough finite values and if both f_1 and f_2 are analytical one-to-one functions of finite (low) order.

Rather than discussing this conjecture in all necessary generality, it is in the present context preferable to illustrate it with a numerical example. Figure 1 shows the toroidal map (endomporphism) of Eq. (1), with two arbitrary functions f_1 and f_2 inserted and with both ε and δ relatively large. The impression one gains is still that of the second iterate of a solenoid, 'ironed flat'. This behavior is numerically robust.

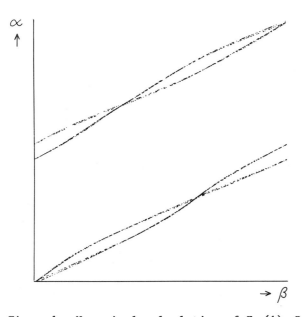

Figure 1.   Numerical calculation of Eq.(1). Both α and β run from zero through 2π (modulo 2π) and hence form a toroidal map. (The terms $2\beta_n$ and $2\alpha_{n+2}$ were also taken modulo 2π.) Parameters: ε = 0.1π, δ = 0.02π. Perturbing functions: f_1 = $\cos^2\alpha_n$ + $\sin^2 3\alpha_n$ + $\sin^3\beta_n$, f_2 = $\sin^3\beta_n$, f_2 = $\sin^3\beta_n$ + $\cos^2 2\alpha_{n+2}$. Initial conditions: $\alpha_1 = 0$, $\beta_1 = 1$. From this initial point, 2000 subsequent iterates are shown. Calculation, with 11-digit accuracy, was performed on a HP 9845B desk-top computer with peripherals.

The main trick which makes the clocks game work (and the conjecture possible) lies with using the same resetting mechanism twice. This way, the inherent instability (amplification of hand angles) applies equally to the second player. The latter at first sight appears to be replaceable by a device that merely more or less accurately preserves the last angular position of the first clock so as to enable it to be reset toward twice its former value. This assumption amounts to replacing the two constants ´2´ in the first line of Eq.(1) by unity. Under this condition, however, no similar conjecture appears possible.

3. A FOUR-VARIABLE EQUATION

The above ´game´ involved three autonomous oscillators or, equivalently, two autonomous oscillators plus one periodic forcing. From the theory of periodically forced two-variable oscillators it is known that every such system can be written as an autonomous three-variable system (cf. Rössler, 1979b, for two appropriate conversion formulas between polar and Cartesian coordinates). This means that in the present case no more than 5 variables are needed. A further reduction to 4 is possible in the present symmetrical situation by applying the same principle (of recovering an independently usable ´third´ variable – amplitude from a sinusoidal two-variable oscillator) twice.

The following 4-variable equation is one possible implementation:

$$\dot{x} = 50(z´-x) \cdot f_1(X-0.5) \cdot f_2(Y-1.6) + x(-Y\overline{R} + Xh)$$

$$\dot{y} = 50(w´-y) \cdot f_1(X-0.5) \cdot f_2(Y-1.6) + y(-Y\overline{R} + Xh)$$

$$\dot{z} = 100(x´-z) \cdot f_3(\ Y-1\ ) \cdot f_4(2.3-X) + z(\ X\overline{R} + Yh)$$

$$\dot{w} = 100(y´-w) \cdot f_3(\ Y-1\ ) \cdot f_4(2.3-X) + w(\ X\overline{R} + Yh)$$

$$(2)$$

where $z´ = 0.7(z^2-w^2)$, $w´ = 0.7(2zw)$, $x´ = 0.5(x^2-y^2)$, $y´ = 0.5(2xy)$, $\overline{R} = R-2$, and $h = 3(1-\overline{R}^2)(\overline{R}+0.205X-0.12Y)-\overline{R}$. Three times in this equation, the typical ´amplitude´ variable of a rotation-symmetric oscillator appears: $X = x^2 + y^2$, $Y = z^2+w^2$ and $R = \sqrt{X^2+Y^2}$, the latter (unlike the former) being unsquared. The highest-level oscillator in this nested 3-oscillator system is the one in X and Y (with amplitude R). It produces a relaxation oscillation in the X, Y plane in which either variable assumes small,

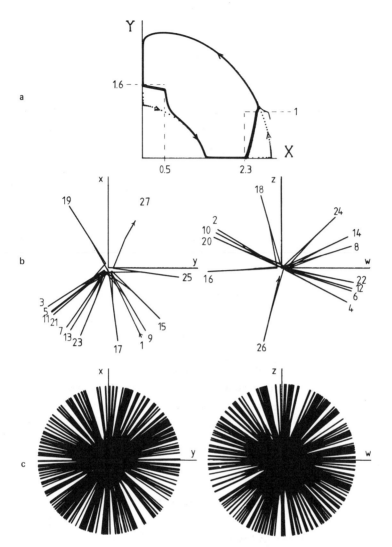

Figure 2. Numerical simulation of Eq.(2). a: State plane of the highest-level suboscillator: amplitudes X and Y of the two ´clocks´ (x,y) and (z,w), respectively. The dotted portions of the overall limit cycle apply when there is no resetting between the two suboscillators (cf. text). b: Two 2-dimensional projections of the whole 4-dimensional flow. Shown are the two ´clock planes´ (x,y) and (z,w). Subsequent angular positions of the two suboscillators (´clock hands´) are numbered. c: The same, continued for a 19 times longer time interval. Axes: 0...3 for X and Y, and 0...2 for x, y, z, w. Initial conditions: x(0) = −1.52, y(0) = 0.748, z(0) = 0.438, w(0) = −0.816; t_{end} = 63.56 (in b). Simulation, using a standard Runge-Kutta-Merson integration routine, was performed on a HP 9845B desk-top computer (cf. Fig. 1).

near-zero values during part of the cycle. This means that the two
0´primary´ suboscillators (in x,y and z,w, respectively) assume small
amplitudes (X and Y, respectively) in alternation. During its small-
amplitude phase, each of these suboscillators (which incidentally does not
oscillate at all - the corresonding frequency terms have been multiplied
by zero and omitted) can easily be reset toward twice the angular position
of the other. This sudden, amplitude-increasing coupling is accomplished
by means of the ´smooth step functions´ f_i. They have the explicit form

$$f(s) = \frac{1}{2} (1 - s(s^2 + 10^{-4})^{-1/2})$$

where s is the argument, being close to unity if s < 0 and close of zero
if s > 0. These functions ´take over´ during part of the cycle (namely,
in two rectangular corners of state space through which the X,Y limit
cycle is passing). The accuracy of the resetting depends mainly on the
smallness of the minimal amplitudes of X and Y, and to a lesser extent on
the smallness of the ´smoothing constant´ (10^{-4} above) in the functions
f_i.

A numerical simulation is shown in Figure 2. Figure 2a shows the
amplitude oscillation (X,Y). The dotted parts apply when the f_i functions
are all zero. Then the rectangles (dashed lines) cut out by the 4
threshold functions are inoperative. Figure 2b shows the two ´clock
faces.´ Subsequent positions of the angles of the two suboscillators have
been marked by consecutive numbers. Comparison with the faces of two real
clocks (each with the usual 12-hour face) shows that the above two-clocks
game is indeed literally implemented in the present 4-variable O.D.E. -
although with a small error. A much longer simulation is shown in the
bottom Figure (Fig. 2c).

It is worth noting that none of the parameters used in Eq.(2) is very
critical. Replacing for example the two constants 0.205 and 10^{-4} by 0.21
and 10^{-3}, respectively, still yields a similar outcome.

4. QUALITATIVE STRUCTURE OF THE SYSTEM

The map (Poincaré cross-section) which applies to the system of
Eq.(2) is not easy to derive. It is a diffeomorphism, since the equation
has Lipschitzian right-hand sides on the one hand, and since there are no
steady states in the region of interest, on the other. The inherent
symmetry suggests a decomposition of the overall map into two (almost)
identical sub-maps, one of these being a cross-section cutting

symmetrically through the middle of the lower curved arc in Figure 2a, the other cutting in the same way through the middle of the upper arc. As it turns out, the resulting submap cannot readily be visualized in three dimensions even though it is an M^3. Specifically, it appears to be a solid (thickened) version of a two-manifold (M^2) that itself, although being closely related to a torus, cannot be smoothly embedded in three dimensions. It is the torus formed by joining together two Klein bottles, after having cut off the ´handle´ of each, in a point-symmetric fashion. (Think of a doubly twisted band closed into a torus.) This torus has two sides (is orientable) but has no interior. The thin (and squeezed and sheared) image of the ´torus´ then lies inside its thick wall, after a two-fold stretching and wrapping. A more straightforward cross-section (an ordinary thickened torus), for which however no simpler half-iterate exists, can be obtained by making a horizontal cut through the middle of the upper arc in Figure 2a and plotting arctan (y/x) vs. arctan (w/z) vs. Y. One then obtains what looks like a 3-dimensional (thickened and diffeomorphic) version of Figure 1. Further numerical studies of Eq.(2) may be worthwhile.

5. DISCUSSION

A four-variable equation has been proposed which was devised in such a way as to possess the second iterate of a solenoid as a cross-section. Even though the actual solenoid is ´sheared´ (a property lacked by Smale´s, (1977) solenoid), the flow on the attractor may still be hyperbolically expanding every-where. A set of parameters that is optimal for numerical studies, and another yielding an appropriate (for example, singular perturbation) limit for analytical studies have yet to be found. At the time being, Eq.(2) is only a ´candidate´ for an axiom A strange attractor in the same sense as the Lorenz equation (Lorenz, 1963) is a candidate for a non-axiom A strange attractor (cf. Guckenheimer, 1976).

The demonstration of differential equations implementing axiom A systems is of a certain mathematical interest. So far, all chaos-generating 3-variable O.D.E.´s with diffeomorphic cross-sections belong to the walking-stick map type (cf. Rössler, 1979c). These systems are not axiom A. While they do generate a chaotic attractor (with a cross-section that is a fractal; cf. Rössler, 1980), this fractal is not ´pure´ in the sense that it in general contains small contracting portions. (An analogue is generated in a solenoid - and Eq.(2) - if the first iterate is allowed to contain a small contracting segment.) These contracting portions (which in general contain a periodic attractor) interestingly do not spoil the geometric shape of the attracting whole object ('first-level attractor') of which they are a part. Partly because of this reason (but mostly because of the numerical inaccessibility of these pockets), chaos-generating systems of this impure (non-axiom A) type nowadays frequently are said to possess a 'strange attractor.' This should nonetheless be avoided since this notion is reserved for axiom A systems and their non-differentiable (mostly Lorenzian, so far) counterparts. The original notion of a strange attractor was, of course, synonymous to an axiom A attractor in the sense of Smale (Ruelle and Takens, 1971).

Diffeomorphism are mathematically simpler than their benignly (in a non-overlapping manner) endomorphic competitors that are only almost everywhere diffeomorphic - like the baker's transformation and the related Lorenz map. To judge from the present example, it seems that strange attractors based on pure diffeomorphisms are necessarily more complicated in terms of the underlying equations. Eq.(2) has 4 variables and many nonlinear terms. This number may be somewhat reducible. Lorenz-type attractors, on the other hand, require no more than two quadratic terms in 3 variables (Lorenz, 1963; Rössler, 1976, 1979a). It is dubious whether this gap can be closed. As to mere chaotic attractors, it is known that a single quadratic term in a 4-variable system suffices even to generate 'higher-order' chaos (with two directions of hyperbolic instability rather than one; Rössler, 1979c; see also Froehling et al., 1981). Nonetheless, the very prototypical nature of attracting axiom A chaos makes it desirable to a look at its 'neighbors' in parameter space. Explicit continuous examples like Eq.(2) may prove useful in making such bifurcation studies feasible.

A preliminary version of the present equation was presented at the 1981 Biophysical Society Meeting (Rössler, 1981).

REFERENCES

1. Froehling, H., J. P. Crutchfield, D. Farmer, N. J. Packard and R. Shaw: On determining the dimension of chaotic flows. Physica D, 3D, 605-617, 1981.
2. Guckenheimer, J.: A strange, strange attractor. In: The Hopf Bifurcation and Its Applications (J. E. Marsden and M. McCracken, eds.), pp. 368-381. New York: Springer-Verlag 1976.

3. Hopf, E.: Ergodentheorie (Ergodic Theory), p. 42. Berlin: Springer-Verlag 1937.

4. Lorenz, E. N.: Deterministic nonperiodic flow. J. Atmos. Sci., 20, 130-141, 1963.

5. Plykin, R. V.: Sources and sinks for A diffeomorphisms. Mat. Sb. 23, 233-253, 1974.

6. Rössler, O. E.: Different types of chaos in two simple differential equations. Z. Naturforsch. a, 31a, 1664-1670, 1976.

7. Rössler, O. E.: Continuous chaos - four prototype equations. Ann. N.Y. Acad. Sci. 316, 376-392, 1979a.

8. Rössler, O. E.: Chaotic oscillations - an example of hyperchaos. In: Nonlinear Oscillations in Biology (F.C. Hoppensteadt, ed.), pp. 141-156. Lectures in Applied Mathematics, Vol. 17. Providence, R.I.: Amer. Math. Society 1979b.

9. Rössler, O. E.: Chaos. In: Structural Stability in Physics (W. Güttinger and H. Eikemeier, eds.), pp. 290-309. Berlin-Heidelberg-New York: Springer-Verlag 1979c.

10. Rössler, O. E.: Chaos and bijections across dimensions. In: New Approaches to Nonlinear Problems in Dynamics (P. H. Holmes, ed.), pp. 477-486. Philadelphia: Society for Industrial and Applied Mathematics 1980.

11. Rössler, O. E.: A continuous axiom A system. Biophys. J. 33, 54a, 1981. (Abstract.)

12. Rössler, O. E. and P. Ortoleva: Strange attractors in 3-variable reaction systems. Springer Lecture Notes in Biomathematics 21, 67-73, 1978.

13. Ruelle, D. and F. Takens: On the nature of turbulence. Commun. Math, Phys. 20, 167-192, 1971.

14. Smale, S.: Differentiable dynamical systems. Bull. Amer. Math. Soc. 73, 747-817, 1967.

15. Smale, S.: Dynamical systems and turbulence. Springer Lecture Notes in Mathematics 615, 48-70, 1977.

16. Williams, R. F.: The bifurcation space of the Lorenz attractor. In: Bifurcation Theory and Applications in Scientific Disciplines (O. Gurel and O. E. Rössler, eds.), Ann. N.Y. Acad. Sci. 316, 393-399, 1979.

PART II

7

Is There Chaos Without Noise?

Ralph H. Abraham

Mathematics Board
University of California
Santa Cruz, California

Dedicated to Mauricio Peixoto.

The chaotic attractor of dynamical systems theory has been widely heralded as a new paradigm for chaotic and turbulent motions in nature. This idea has been strongly supported by the experimental discovery of the chaotic attractor of simulation machines. Is this experimental object an instance of the mathematical model, or an artifact of noise amplification? Here, we establish the existence of this artifact in the forced Van der Pol system, explain how it could account for the experimental chaos of the Lorenz, Rössler, and Shaw systems, describe a critical experiment to distinguish between the noise-amplification and the chaotic attractor models, and propose a new concept of dynamical stability.

1. INTRODUCTION

The chaotic attractor of mathematical theory began with Birkhoff in 1916. The chaotic attractor of simulation experiment arrived with Lorenz in 1962. (See Abraham and Shaw, 1983a, for historical details.) The identification of these two objects has not yet succeeded, despite many attempts during the past twenty years. Of course, everyone (including myself) expects this to happen soon (see Hirsch, 1983). Nevertheless,

there is an important reason to consider the loathsome alternative: the
quasi-periodic paradox.

2. THE QUASI-PERIODIC PARADOX

The attractive, invariant 2-torus is ubiquitous in dynamical systems. It
occurs, for example, in the "main sequence" of bifurcations: static
attractor to periodic attractor by a Hopf bifurcation, to an attractive
invariant 2-torus (AIT) by a Neimark bifurcation. It is always found in
forced oscillators,such as the Rayleigh or Van der Pol. According to
Peixoto´s Theorem on the open genericity of structural stability for flows
on the 2-torus, the restricted flow on the AIT must almost always be a
braid of periodic attractors. Thus, the power spectrum of one coordinate
of a typical trajectory would reveal fundamental frequencies of two modes
of oscillation, rationally related. However, most of the time,
experimentalists observe not braids (rationally related frequencies) but
quasi-periodic motions (apparently irrationally related frequencies).
That is the quasi-periodic paradox. More than one scientist has lost
faith in mathematics because of the ubiquity of this illegal motion in the
natural world. In fact, in the forced Van der Pol system, quasi-periodic
motion persists over most of the parameter space (see Abraham and Scott,
1983). We now present three competing explanations of this paradox.

3. THE THICK BIFURCATION MODEL

This scheme is due to Sotomayer (1974). We suppose that the dynamical
system in question has one loose parameter. Thus, we are observing not a
single generic dynamical system on a single AIT, but a generic arc of
dynamical systems on a moving AIT. Therefore the braid bifurcations, at
which one braid of periodic attractors changes to another (the ratio of
frequences changes from one rational to another nearby rational), occur
very frequently along this arc. In fact, Herman (1979) showed that the
probability of bifurcation may be close to one. Thus, most of the time,
quasi-periodic motions will be observed. This explanation of the quasi-
periodic paradox is favored by mathematicians.

4. THE NOISE MODULATION MODEL

On the other hand, this one is favored by physicists (Ueda and Akamatsu,
1981; Shaw, 1981). Here we assume that the mathematical model is a single

generic dynamical system, with a braid of periodic attractors on an AIT. But the experimental system simulating it has noisy imperfections. For example, the function generator providing the forcing voltage to an analog computer has low-level noise in its power spectrum. This noise, if its amplitude is sufficient, will cause the trajectory to leap from one basin of attraction, on the AIT, to another. The smallest distance form a periodic attractor to its separatrix (a periodic saddle) determines the critical amplitude of noise sufficient for quasi-periodic motion.

5. THE CHAOTIC ATTRACTOR MODEL

This rationale is a compromise of the two preceding ones. While admitting that noise modulation occurs in the simulation device, we suppose that there is a dynamical system model for the device, including its noise. At this point, we resort to the fact that chaotic attractors are known to exist in mathematical theory, even if we do not yet know whether the Lorenz system (for example) has one, or not. Thus, a mathematical model for the simulation device may be regarded as a coupling (generic perturbation) of the Cartesian product of an AIT (in three dimensions) and an unknown system with a chaotic structure (in three dimensions or more). Thus, our observation of the attractor in three dimensions is a projection of the actual chaotic attractor in six dimensions or more. Hence, there is no conflict with Peixoto's theorem.

This obviously models the noise modulation scheme. It may be applied to the thick bifurcation scheme as follows. Make a dynamical model for the ambient noise in the single loose parameter. Serially couple this model to the generic one, by selecting one coordinate (or some other real-valued function of the state space of the chaotic model) to control the loose parameter. The resulting coupled system will be quasi-periodic most of the time.

6. APPLICATION TO EXPERIMENTAL CHAOS

We may take the best known chaotic flows of experiment (Lorenz, 1962; Rössler, 1971; Shaw, 1981) as examples. For the sake of discussion, we suppose (banish the thought) that these systems do not contain a mathematical chaotic attractor. Then their apparently chaotic behavior may, like the quasi-periodic paradox, be explained by the noise modulation (and related) schemes. For in each of these cases, the subject dynamical system is known to have an attractive invariant set (AIS) which behaves like the AIT of the quasi-periodic paradox. For example, the AIS of the Lorenz system is the outset structure of its three saddle points (Lorenz, 1963; Abraham and Shaw, 1983b). The AIS of the Rössler system is the outset, a Möbius band, of its fundamental limit cycle (Rössler, 1979; Abraham and Shaw, 1983a). Finally, the AIS of the Shaw system is the usual AIT of a forced oscillator (Shaw, 1981; Abraham and Shaw, 1983a; Abraham and Scott, 1983).

And thus our question: is there chaos without noise? That is: is there a chaotic attractor in these particular three-dimensional systems (Lorenz, Rössler, Shaw) without modulating noise?

7. A CRITICAL EXPERIMENT

We do not know the answer to this question. It could be yes for the Shaw system, for example, and no for the Lorenz and Rössler systems. We do know that it is no for the quasi-periodic motion on an AIT. So we imagine there could be an experimental test for noise modulation; that is, a procedure to exclude noise modulation as a model for chaotic behavior of experimental systems. Here is a rough sketch of one such procedure.

Suppose a system is given with a single control parameter, which has a bifurcation to chaos at one bifurcation value of the control. Then, if the chaotic behavior is due to noise modulation, the bifurcation to chaos will depend on the amplitude of the noise. For example if the noise in the forced oscillator system is reduced, the appearance of quasi-periodic motion on the AIT will occur for a higher value of the amplitude of the forcing oscillation. Let noise ratio denote the ratio of power in the continuous part of the power spectrum of the simulated system, to that of simulation device at rest. Then the noise ratio, as a function of the bifurcation parameter, could discriminate between noise-modulated, versus truly chaotic, behavior.

8. DYNAMICAL STABILITY

The ubiquity of structurally unstable motions, in conflict with Peixoto's Theorem in the AIT context, suggests that structural stability is not an appropriate concept for experimental systems. Here, we suggest an alternative, very much in the spirit of Ueda and Akamatsu (1981). Suppose our dynamical system, for simplicity, has a single attractor, and upon perturbation , it still has a single attractor. In fact, let us suppose it is structurally stable, so that the attractor is not significantly changed by any small, static perturbation. Supposing the system depends on control parameters, let us not serially couple the output of another (possibly chaotic) system to these controls. Then the original system is dynamically stable if its attractor is not significantly changed by this dynamical perturbation, provided it is sufficiently small.

For example, the braided attractors on an AIT are structurally stable, in the static, classical sense, according to Peixoto's Theorem. But they are not dynamically stable, because any amount of dynamical perturbation (even periodic perturbation) may produce a chaotic attractor.

9. CONCLUSION

The formulation of the chaotic attractor model for noise modulation solves the quasi-periodic paradox. It may solve a chaotic attractor paradox, if there is one, in specific systems such as the Lorenz system. We still do not know if there is chaos without noise in these systems, or not. But noise ratio experiments may shed some light on this question. This is a special case of a more general question: are these systems dynamically stable? Here we may hazard a conjecture: all natural systems are dynamically stable. In fact, we will probably evolve the definition of stability until this conjecture becomes true.

REFERENCES

1. Abraham, R. H. and K. A. Scott, 1983. Chaostrophes of forced Van der Pol systems. This volume.
2. Abraham, R. H. and C. D. Shaw, 1983a. Dynamics, the Geometry of Behavior. Part Two: Chaotic Behavior. Aerial, Santa Cruz, CA.
3. Abraham, R. H. and C. D. Shaw, 1983b. The outstructure of the Lorenz attractor. This volume.
4. Herman, M., 1979. Sur la conjugaison différentiable des difféomorphismes du circle à des rotations. Publ. Math. I.H.E.S., 49.
5. Hirsch, M. W., 1983. The chaos of dynamical systems. This volume.
6. Lorenz, E., 1963. Deterministic nonperiodic flow. J. Atmos. Sci. 20: 130-141.
7. Rössler, O. E., 1976. An equation for continuous chaos. Phys. Lett. 57A: 397-398.
8. Shaw, R. S., 1981. Strange attractors, chaotic behavior, and information flows. J. Naturforsch. 36a: 80.
9. Sotomayer, J., 1974. Generic one-parameter families of vector fields in two-dimensional manifolds. Publ. Math. I.H.E.S. 43: 5-46.
10. Ueda, Y. and N. Akamatsu, 1981. Chaotically transitional phenomena in the forced negative-resistance oscillator. IEE Trans. CAS-20: 217-224.

8

Chaostrophes of Forced Van der Pol Systems

Ralph H. Abraham

Mathematics Board
University of California
Santa Cruz, California

Katherine A. Scott

Computer Center
University of California
Santa Cruz, California

Dedicated to Chihiro Hayashi.

In response to a recent conjecture, we explored two systems in analog simulation, in search of the blue bagel chaostrophe--in which a chaotic bagel attractor disappears into the blue. We found an abundance of these bifurcations in the forced Van der Pol systems.

1. INTRODUCTION

In a recent conjecture (Abraham, 1983a, Section A4 and Fig. 3) a blue bagel chaostrophe is proposed to exist in the forced Van der Pol system. This conjecture was based on the discovery of the chaotic bagel attractor in systems of this type by R. Shaw (1981). Our idea was to study these same systems in analog simulation, turning the knobs until the bagel attractor collided with a homoclinic tangle, its separatrix, in a mutual annihilation. We are grateful to Rob Shaw for sharing his lab with us for these experiments. What we found, more complicated than expected due to period doubling bifurcations, is presented here. The background of this entire cycle of ideas is a drawing by Hayashi, Ueda and Kawakami (1970) gracing the cover of Hayashi´s volume of selected papers (1975, p. 186).

This drawing clearly shows an AIT (attractive, invariant torus) within a
homoclinic tangle, occurring in a forced, conservative, Duffing system.

2. ACCELERATION FORCING

First we explored the conventionally forced Van der Pol system, in the
form:

$$\dot{x} = ky + \mu x(a - y^2) + A\sin\theta$$
$$\dot{y} = -x$$
$$\dot{\theta} = 2\pi F$$

The fixed parameters, $a = k = 9$, $\mu = 32$, were chosen by twisting knobs and
looking for likely attractors in the strobe plane. The analog setup in
Rob Shaw's lab, used for this work, conveniently includes a strobe pulse
and storage scope for direct observations of the of the Poincaré section.
Amid a sea of complex bifurcations, we selected a simple arc, shown in
Fig. 3. Here A = 0.420, fixed, while F varies from the frequency of the
periodic attractor of the unforced system, 4.2 Hz, to double that, or 8.4
Hz. In Fig. 3 we see:
Two views of the fundamental oscillation--
    A. Full periodic attractor, F = 4.20 Hz.
    B. Strobed periodic attractor, F = 4.20.
Subtle bifurcation to chaotic torus--
    C. Full baby torus, F = 6.60
    D. Strobed baby torus, F = 6.60
    E. Strobed large torus, F = 7.05
    F. Two strobed phases, 0 and π, F = 7.05
    G. Strobed torus, F = 8.00.
Catastrophic bifurcation back to the fundamental--
    H. Strobed periodic attractor, F = 8.15.
These two bifurcations are apparently formed by a nearby homoclinic
tangle. However, we did not observe it. The large attractor, an AIT,
appears quasi-periodic because of noise modulation (see Abraham, 1983b).

3. VELOCITY FORCING

To observe analogous behavior in a chaotic system, we moved the driving
oscillator form the acceleration equation to the velocity, following R.
Shaw (1981). Thus, we explored the system

$$\dot{x} = ky + \mu x(a - y^2)$$

$$\dot{y} = -x + A\sin\theta$$

$$\dot{\theta} = 2\pi F$$

with the same parameters, $a = k = 9$, $\mu = 32$, $A = 0.420$, and F between 4.2
and 10.0 Hz. The results are shown in Fig. 4:
The fundamental oscillation--

    A. Strobed periodic attractor, F = 6.15.
Chaostrophic bifurcation to a bagel--

    B. Strobed bagel, F = 6.20.
This continues over a large range--

    C. Strobed bagel, F = 9.20.
and chaostrophically collapses again to the fundamental,

    D. Strobed periodic attractor, F = 9.25.

Again, these two catastrophic bifurcations are apparently related to
nearby homoclinic saddles, but we did not observe them. In particular,
the bagel in Fig. 4C clearly shows extensive dwell at the top and bottom,
suggesting the locations of the nearby, invisible, saddle orbit of period
two.

    In Fig. 3, we conjecture a rough idea of the nearby homoclinic
tangles, following the inspiration of Hayashi, Ueda and Kawakami (1969, p.
251): For the conventionally forced system of the preceding section,
forced at about the fundamental frequency--

    A. Corresponding to Fig. 3D, the baby torus in a homoclinic nest.
And at the end of its regime, at about double the fundamental frequency--

    B. Corresponding to Fig. 3G, the great torus in a homoclinic box.
And for the velocity forced system of this section, at about the
fundamental frequency--

    C. Corresponding to Fig. 4B, the chaotic bagel in a nest-- and again
at about double the fundamental frequency--

    D. Corresponding to Fig. 4C, the chaotic bagel in a homoclinic box.

4. BIFURCATION CONJECTURES

Without further experimental work, we may only guess at the bifurcation
sequences behind our observations, shown in Figs. 1 and 2. Here, we
record a few guesses suggested by the observations. Consider first the
bifurcation, shown in Poincaré section from Fig. 1B to 1D. As the

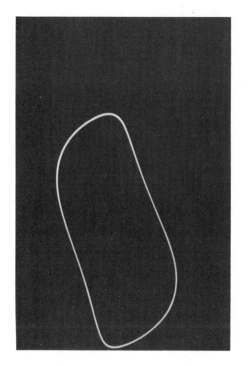

A

B

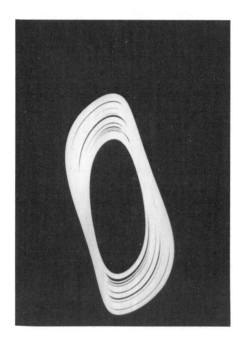

C

D

FIG. 1. AIT bifurcations observed in Van der Pol´s system.

A. F = 4.20 Hz., fundamental.

B. F = 4.20 Hz., strobed.

C. F = 6.60 Hz., baby AIT.

D. F = 6.60 Hz., strobed.

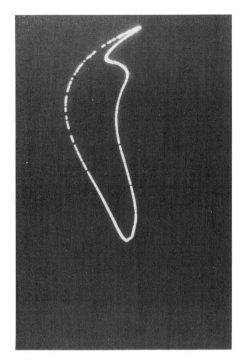

E

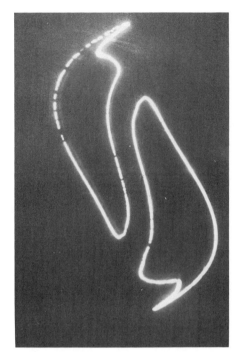

F

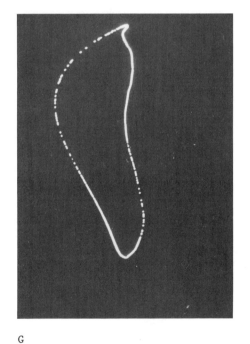

G

H

FIG. 1 (cont.)

E. F = 7.05 Hz., torus.

F. F = 7.05 Hz., opposite phases.

G. F = 8.00 Hz., blue torus.

H. F = 8.15 Hz., fundamental.

A

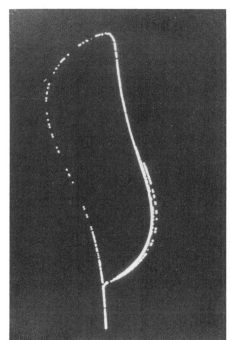

B

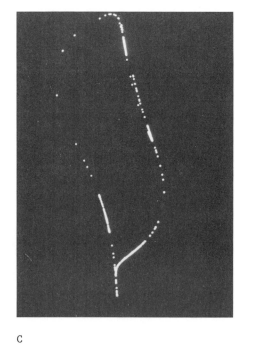

C

D

FIG. 2. Bagel bifurcations observed in Shaw's system.

A. F = 6.15 Hz., fundamental.

B. F = 6.20 Hz., blue bagel.

C. F = 9.20 Hz., blue bagel.

D. F = 9.25 Hz., fundamental.

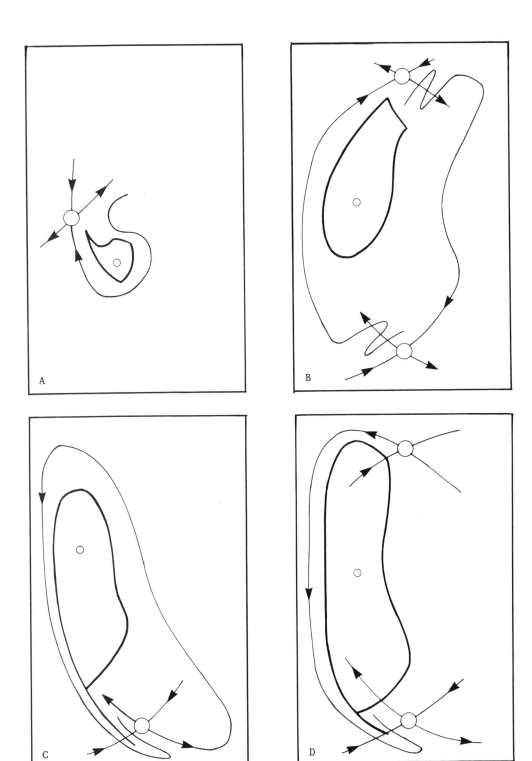

FIG. 3. Conjectured Homoclinic tangles.

A. Like 1D.

B. Like 1G.

C. Like 2B.

D. Like 2C.

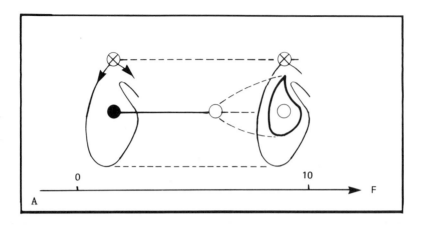

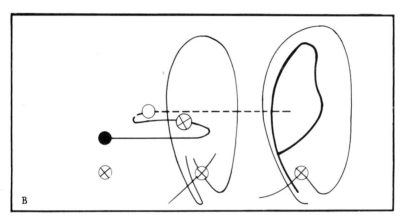

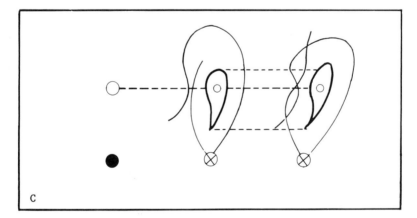

FIG. 4 Conjectured bifurcation diagrams.

A. Hopf in a nest.

B. Blue torus or bagel.

C. Captive balloon.

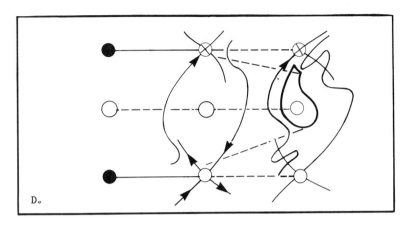

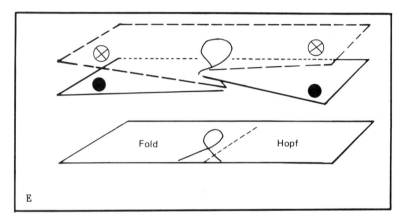

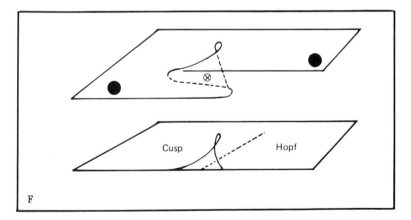

FIG.4 (cont.)

D. Subharmonic balloon.

E. Poppyseed.

F. Blueberry.

teardrop shaped section of an AIT was observed to grow from a point as the driving frequency, F, increased, a Neimark bifurcation within a homoclinic tangle is an obvious guess, as shown in Fig. 4A.

However, in the corresponding bifurcation in Shaw's system, from Fig. 2A to 2B, the chaotic bagel springs forth fully grown. So we conjecture here a blue bagel catastrophe, as shown in Fig. 4B. The disappearance of the fundamental periodic attractor, due (in this proposed model) to a static annihilation (saddle-node) catastrophe as the frequency is increased, drops the trajectory onto the bagel, recently appeared out of the blue, through the formation of a homoclinic tangle by the lower periodic saddle. We could apply this model to the preceding bifurcation as well.

These models are consistent with the well established theory of generic arcs of diffeomorphisms. But based on the observations, we would prefer a model in which the fundamental attractor destabilizes into a periodic saddle with a homoclinic tangle, within which an AIT or a bagel simultaneously forms out of the blue, as shown in Fig. 4C. This captive balloon catastrophe is theoretically unlikely, except at a bifurcation of codimension two. Nevertheless, we seem to observe this repeatedly in the analog simulations. This is an unsolved paradox at present, and deserves further study. Suitable bifurcations of codimension three, containing all three proposed arcs (4A, 4B, 4C) are shown in Figs. 4E and 4F. A third control parameter, not shown, creates a saddle connection. The bifurcations at the higher forcing frequencies, from Fig. 1H to 1G and from 2D to 2C, appear to be subharmonic (period two) versions of the captive balloon catastrophe, as shown in Fig. 4D.

5. CONCLUSION

The occurrence of toral and bagel chaostrophes in forced Van der Pol systems is established. It remains to draw the surrounding homoclinic tangles, in the wonderful style of Hayashi, by actual simulation instead of fantasy. But that is very difficult. The ubiquitous coincidence of two bifurcation events, called here the captive balloon catastrophe, also suggests further work, in search of a poppyseed bifurcation.

BIBLIOGRAPHY

Abraham, R. H., 1983a, Chaostrophe, intermittency, and noise. This volume.

Abraham, R. H., 1983b, Is there chaos without noise? This volume.

Abraham, R. H. and C. D. Shaw, 1983, Dynamics, The Geometry of Behavior, Part Two: Chaotic Behavior, Aerial, Santa Cruz, CA.

Hayashi, C., 1975, Selected papers, Nippon, Kyoto.

Hayashi, C., Y. Ueda and H. Kawakami, 1969, Transformation theory as applied to the solutions of non-linear differential equations of the second order, Int. J. Non-Linear Mechanics, 4: 235-255.

Hayashi, C., Y. Ueda, and H. Kawakami, 1970, Periodic solutions to Duffing´s equation with reference to doubly asymptotic solutions, Proc. 5th Int. Conf. on Nonlinear Oscillations, 2: 507-521.

Langford, W. F., 1982, Chaotic dynamics in the unfoldings of degenerate bifurcations, Proc. Int. Symp. Applied Mathematics and Information Science, Kyoto Univ.

Tomita, K., 1982, Chaotic response of non linear oscillators, Phys. Rep., 86: 114-167.

9

Numerical Solution of the Lorenz Equations with Spatial Inhomogeneity

M.E. Alexander
Institute of Computer Science
University of Guelph
Guelph, Ontario, Canada

J. Brindley and I.M. Moroz *
School of Mathematics
University of Leeds
Leeds, England

The classical Lorenz equations describe the evolution in time, but not in space, of a variety of physical systems possessing weakly nonlinear behavior: for example, baroclinic waves in the earth's atmosphere. When these equations are generalized to include spatial structure, there arises the question of the stability of wave trains (previously modelled by the classical equations) to spatial perturbations. This paper discusses a numerical scheme for the solution of the partial differential equations that generalize the classical Lorenz equations, and whether the numerical solution is a meaningful approximation to the corresponding analytic one. Some insight into both anaytical and numerical solutions is achieved by considering a simple system of two interacting spatial Fourier modes. It is concluded that the Lorenz equations, and their generalization, are not viable models for describing physical behavior.

INTRODUCTION

It has been shown recently (Gibbon and McGuinness 1980, Brindley and Moroz 1980) that the equations governing the weakly nonlinear behaviour of

---

*Present address: School of Mathematics and Physics, University of East Anglia, Norfolk, England

unstable baroclinic waves in a two-layer or "Eady" type model of baroclinic instability are formally identical to the Lorenz equations (Lorenz 1963) of a simple model for thermal convection. Detailed discussion of the models and derivation may be found in Pedlosky (1979) and Moroz and Brindley (1983).

The Lorenz equations occur when the disturbance is assumed to comprise a single mode only; the behaviour described is that of an infinite wave train having a particular wave number K. In an effort to construct a more realistic model Brindley and Moroz (1983) considered the nonlinear evolution of a wave packet comprising a narrow band of wave-lengths centred on K. For the values of dissipation leading to the Lorenz equations the most unstable wave number on linear theory is K = 0, but the analysis was nevertheless pivoted about a value of K of O(1) both on physical grounds and in order to justify a multiple scales approach in which amplitude modulation occurred on a scale large compared with K^{-1}. Pedlosky (1981) has examined the case in which K≪1, but non-zero.

The resulting amplitude evolution equations are now <u>partial</u> differential equations, differing from the Lorenz <u>ordinary</u> differential equations in the occurrence of a spatial derivative; their coefficients may be real or complex depending on the absence or presence of dispersion in the model.

In this paper we examine numerical solutions to these "spatial" Lorenz equations for a range of parameter values, and discuss the relevance of the results to the behaviour of real systems.

1. GENERAL NUMERICAL METHOD

A set of computer programmes was developed for solving a general system of K coupled nonlinear partial differential equations in one space variable, ξ, and time, t, of the form

$$u_t(\xi,t) = L(u,\xi,t) \qquad 0 < \xi < 1 \qquad (1.1)$$

where L is a (nonlinear) differential operator with space derivatives of order $\leq$ m and $u = (u^1, u^2, ..., u^q) \in \mathbb{R}^q$. Periodic boundary conditions

$$\frac{\partial^i U(0,t)}{\partial \xi^i} = \frac{\partial^i U(1,t)}{\partial \xi^i} \qquad i = 0, 1, ..., (m-1) \qquad (1.2)$$

were imposed, reflecting the physical origin of the equations as representing wave motions in a finite annulus. The numerical method employed for solving these equations was a Galerkin scheme, having identical trial and test spaces consisting of quadratic B-splines, with compact support on an interval of width 3 mesh intervals and belonging to class $C^1[0,1]$. With this basis, ξ - derivatives of order $m\leq4$ can be handled. Furthermore, a uniform mesh was used throughout.

The computer programme runs in three consecutive stages:

(i)   Preparation of an input data file, specifying parameter values and initial conditions, as well as the number of meshpoints N, the size and number of timesteps and error tolerances for the time integrations.

(ii)   Integration of the Galerkin equations derived from (1.1) and (1.2), using input data from step (i). The equations are integrated forward in time using a Crank-Nicholson scheme with Newton iteration to solve the nonlinear equations. The solutions are second-order accurate in time, and the step size is automatically controlled by the programme according to prescribed error bounds. Results at each timestep are output to a file; and at the end of the programme run the input file is updated with the current solution vector, so that it can provide the "initial" conditions for a subsequent run.

(iii)   The data in the output file are displayed graphically by means of a solution surface plot. The solutions data can be either the components $u^{\alpha}(1 < \alpha < q)$, or some transformation v^{α} of these components (for example, if K=2 and u is considered as a complex variable, we could choose $v^1 = |u|$, $v^2 = \arg(u)$). For a given α, the solution is then displayed in (ξ, t, v^{α}) - space in variously chosen projections.

Evaluations of the function and Jacobian (required for the Newton scheme), as well as the transformations: $u \rightarrow v$ of the solution vector, are performed by three subroutines supplied by the user. All other parts of the programme are general-purpose routines that can be kept in a library. The most time-consuming part of the computation comes from the Newton iteration scheme to solve the nonlinear equations at each time step. The Jacobian is re-evaluated only after every 20 iterations, or whenever the timestep used by the programme is altered. This requires the LU

decompostion of a banded matrix with borders, the bandwidth b being
determined by the order of the spline basis and the number of components q
of the solutions: for quadratic splines and q = 5 (corresponding to the
complex Lorenz equations below), b = 12. Both the decomposition and back-
substitution required by the iteration scheme can be performed in O(N)
operations, where N is the number of meshpoints.

2. THE COMPLEX LORENZ EQUATIONS

The equations that were solved by the method described above are a
generalization of the well-known Lorenz equations (Lorenz, 1963; Brindley
and Moroz, 1980) to include a spatial derivative term (Alexander, Brindley
and Moroz, 1982 – hereafter ABM) and complex parameters (Fowler et al.,
1982)

$$X_t = \sigma(Y-X)$$
$$Y_t = (r-Z)X - (1-ie)Y - i\varepsilon X_\xi \qquad (2.1)$$
$$Z_t = -bZ + \frac{1}{2}(X^*Y + XY^*)$$

where $r = r_1 + ir_2$, X and Y are complex, and $\sigma, e, \varepsilon, b$ and Z are real; and **b**
and σ are positive and (*) denotes complex conjugate. The special case e
$= r_2 = 0$ is analogous to the ordinary Lorenz equations, in real X, Y and
Z, apart from the spatial derivative term, (see ABM). The "complex"
Lorenz system without the spatial term – $i\varepsilon X_\xi$ (so that (2.1) reduces to a
set of ordinary differential equations in t), has been investigated by
Fowler et al. (1982), who found behaviour entirely different to the real
case, including the existence of limit cycles.

All these equations describe the weakly nonlinear evolution of a
variety of physical systems with weak dissipation. In the real case,
dispersion is not present, while in the complex case it is included. The
ordinary equations describe the time evolution of infinite wave trains.
However, a more realistic model, including spatial derivative terms as in
(2.1), would describe the evolution of wave packets (Moroz, 1981).
Moreover, it is likely that an infinite wave train could become unstable
to spatial perturbations (Lange and Newell, 1974). This raises two
points: firstly, the analysis of Equations (2.1) may reveal that, for
certain ranges of wavelength of the spatial perturbation, the solution
becomes (linearly) unstable; while outside these ranges it decays or is
stable. Secondly, the analysis of the numerical method itself may

likewise reveal instabilities; and thus it is important to be able to know under what conditions the method goes unstable and how this relates to the behaviour expected from a continuous system described by Equations (2.1).

Since X and Y are complex and Z is real, the set of equations (2.1) are of fifth order. The Galerkin method was implemented by expressing X and Y in terms of their real and imaginary parts: $X = X^r + iX^i$; $Y = Y^r + iY^i$. The Galerkin representation in a B-spline basis $\{\phi_j\}$ for each component of (2.1) is, for N meshpoints,

$$X^r(\xi,t) = \sum_{j=0}^{N-1} x_j^r(t)\phi_j(\xi), \text{ etc.} \tag{2.2}$$

where each ϕ_j is a piecewise polynomial of degree $(\mu - 1)$, belonging to the class $C^{\mu-2}[0,1]$, and is defined by

$$\phi_j(\xi) = \Phi_\mu(\frac{\xi}{h} - j) \tag{2.3}$$

where $h \equiv 1/N$ is the mesh interval, and the function Φ_μ can be derived by a $(\mu-1)$ – fold convolution of the characteristic function:

$$\chi(y) = 1, \qquad |y| \leq \frac{1}{2} \tag{2.4}$$

$$= 0, \qquad |y| > \frac{1}{2}$$

(Schoenberg, 1946), so that, for example, for the quadratic B-splines,

$$\Phi_3(y) = \begin{cases} \frac{1}{2}(y + \frac{3}{2})^2, & -\frac{3}{2} < y \leq -\frac{1}{2} \\[2mm] \frac{3}{4} - y^2, & -\frac{1}{2} < y \leq \frac{1}{2} \\[2mm] \frac{1}{2}(\frac{3}{2} - y)^2, & \frac{1}{2} < y \leq \frac{3}{2} \\[2mm] 0, & |y| > \frac{3}{2} \end{cases} \tag{2.5}$$

The unknown functions $\{x_j^r(t), i = 0, 1, ..., (N-1)\}$ in Equation (2.2) are found by solving the Galerkin equations

$$(\phi_i, U_t^\alpha - L(U^\alpha,.,t)) = 0 \qquad (0 \leq i \leq (N-1); \ 1 \leq \alpha \leq q = 5) \tag{2.6}$$

where the inner product (,) is defined (for real functions f and g) by

$$(f,g) = \int_{-\infty}^{+\infty} f(\xi)\ g(\xi)\ d\xi \tag{2.7}$$

The initial conditions for (2.6) are found by interpolating the initial function at the meshpoints, and solving the resulting linear equations for $\{x_j^r(0)\}$, etc. We shall introduce the following notations:

$$A_j = h^{-1}(\phi_0, \phi_j); \quad B_j = (\phi_0^{\prime}, \phi_j); \quad C_{j\ell} = h^{-1}(\phi_0, \phi_j\ \phi_\ell) \tag{2.8}$$

for $j,\ell = 0, \pm 1, \pm 2, \ldots$. Then, with a little computation we can see that the A_j's, B_j's and $C_{j\ell}$'s are not dependent on h. With the aid of (2.8) we can express the Galerkin equations corresponding to Equations (2.1) in the form

$$A_{k-j}\ \dot{x}_j^r = \sigma A_{k-j}\ (y_j^r - x_j^r)$$

$$A_{k-j}\ \dot{x}_j^i = \sigma A_{k-j}(y_j^i - x_j^i)$$

$$A_{k-j}\ \dot{y}_j^r = A_{k-j}(r_1 x_j^r - r_2 x_j^i - y_j^r - ey_j^i)$$

$$+ \frac{\varepsilon}{h}\ B_{k-j}\ x_j^i - C_{k-j,K-\ell}\ x_j^r\ z_\ell \tag{2.9}$$

$$A_{k-j}\ \dot{y}_j^i = A_{k-j}\ (r_1 x_j^i + r_2 x_j^r - y_j^i + ey_j^r)$$

$$- \frac{\varepsilon}{h}\ B_{k-j}\ x_j^r - C_{k-j,k-\ell}\ x_j^i\ z_\ell$$

$$A_{k-j}\ \dot{z}_j = -bA_{k-j}\ z_j + C_{k-j,k-\ell}\ (x_j^r y_\ell^r + x_j^i y_\ell^i)$$

where summation is over the repeated indices j and ℓ and, to ensure the periodicity of the solution, we define $x_{j+N}^r(t) = x_j^r(t)$, for $j = 0,1,\ldots,(N-1)$. The coefficients $C_{j\ell}$ of the nonlinear terms in Equations (2.9) have the property

$$\sum_\ell C_{j\ell} = A_j \tag{2.10}$$

which is a consequence of definition (2.8) and the property (e.g. Watson, 1979)

$$\sum_\ell \Phi_\ell(\xi) = 1 \tag{2.11}$$

3. ACCURACY AND STABILITY OF THE NUMERICAL METHOD

We shall investigate the convergence and stability of the solutions of the semi-discrete Galerkin equations (2.9) by means of Fourier analysis (e.g., Thomée, 1973; Alexander and Morris, 1979). In order to apply this method, it is necessary to linearize the equations, replacing Z_j in the equations for $\dot{y}_k^r$ and $\dot{y}_k^i$ by a spatially – averaged value $\bar{Z}$, say, so that by property (2.10),

$$C_{k-j,k-\ell}\; x_j z_\ell \;\dashrightarrow\; \bar{Z}\; A_{k-j}\; x_j \qquad (3.1)$$

This results in Z being effectively decoupled from the x and y equations, as regards spatial behaviour, and whatever procedure is used to linearize the last of Equations (2.9) will not affect the discussion that follows. In applying this linearization procedure, we are assuming that the system (2.9) does, in fact, possess a solution for some range of parameter values; and furthermore, that, in a local sense, the solution of the linearized system is close to that of the nonlinear system (2.9).

Now, the equations (2.9) are of the form of discrete convolutions:

$$(f*g)_k \equiv \sum_\ell f_\ell g_{k-\ell} \text{ where } k,\ell = 0,\ \pm 1\ 0,\ \pm 1,\ \pm 2,\ldots \qquad (3.2)$$

so that, if we use the discrete Fourier transform

$$\hat{f}(\theta) = \sum f_\ell e^{-i\ell\theta} \qquad (3.3)$$

and the property

$$(\widehat{f*g}) = \hat{f}\hat{g} \qquad (3.4)$$

these equations can be reduced to 5 ordinary differential equations in time, in which θ appears only as a parameter. Since $\hat{f}$ is complex even when $\{f_j\}$ real, it is easier to apply this method to the equivalent set of complex Galerkin equations

$$A_{k-j}\; \dot{x}_j = \sigma A_{k-j}(y_j - x_j)$$

$$A_{k-j}\; \dot{y}_j = A_{k-j}(rx_j - (1-ie)y_j) - i\frac{\varepsilon}{h} B_{k-j}x_j$$
$$-C_{k-j,k-\ell}x_j z_\ell \qquad (3.5)$$

$$A_{k-j}\; \dot{z}_j = -bA_{k-j}z_j + C_{k-j,k-\ell}(x_j y_\ell^* + x_j^* y_\ell)$$

in which only the last equation is real. If we define

$$g_0(\theta) = \Sigma_\ell A_\ell e^{-i\ell\theta}; \qquad g_1(\theta) = -i\Sigma B_\ell e^{-i\ell\theta} \qquad (3.6)$$

then g_0 and g_1 are real functions and the Fourier transform of the linearized form of the first two equations of (3.5) yields

$$g_0 \frac{\partial \hat{x}}{\partial t} (\theta,t) = \sigma g_0(\hat{y}-\hat{x}) \qquad (3.7)$$

$$g_0 \frac{\partial \hat{y}}{\partial t} (\theta,t) = g_0(r\hat{x}-(1-ie)\hat{y}) + \frac{\varepsilon}{h} g_1 \hat{x} - g_0 \overline{Z\hat{x}} \qquad (3.8)$$

while the equation for $\hat{z}$ can be written

$$g_0 \frac{\partial \hat{z}}{\partial t} (\theta,t) = g_0(-b\hat{z} + Q(x_j,y_j,\theta)) \qquad (3.9)$$

where Q is a (real) quadratic form in x_j and y_j whose linearization need not be specified.

In order to make further progress, we linearize (3.8) in time by setting $\overline{z}$ = constant, and define

$$K(\theta,h) = \frac{1}{h} \frac{g_1(\theta)}{g_0(\theta)} \qquad (3.10)$$

and the matrix

$$\pi(K) = \begin{pmatrix} -\sigma & \sigma \\ (r-\overline{Z}+\varepsilon K) & -(1-ie) \end{pmatrix} \qquad (3.11)$$

If, now, we represent a linearization of Q by

$$Q = a_1\hat{x} + a_2\hat{y} + a_1^*\hat{x}^* + a_2^*\hat{y}^* \qquad (3.12)$$

then the full set of linearized equations for (3.7) − (3.9) may be written

$$\frac{\partial}{\partial t} \begin{vmatrix} \hat{x} \\ \hat{y} \\ \hat{x}^* \\ \hat{y}^* \\ \hat{z} \end{vmatrix} = \begin{vmatrix} \pi & 0 & 0 \\ 0 & \pi^* & 0 \\ a_1 \quad a_2 & a_1^* \quad a_2^* & -b \end{vmatrix} \begin{vmatrix} \hat{x} \\ \hat{y} \\ \hat{x}^* \\ \hat{y}^* \\ \hat{z} \end{vmatrix} \qquad (3.13)$$

Following Thomée (1973), the linearized finite difference scheme (3.5) (with Z_j replaced by $\bar{z}$ in the second equation) is accurate of order p if

$$\pi(\theta,h) = \begin{pmatrix} -\sigma & \sigma \\ (r-\bar{Z}+\varepsilon\frac{\theta}{h}) & -(1-ie) \end{pmatrix} + \frac{1}{h} 0(\theta^{p+1}) \quad \text{as } \theta \to 0 \qquad (3.14)$$

Now, since

$$g_k(\theta) = \theta^k \hat{\Phi}_\mu^2 + R_k(\theta), \quad k = 0,1 \qquad (3.15)$$

where

$$\hat{\Phi}_\mu = (\frac{\sin\frac{1}{2}\theta}{\frac{1}{2}\theta})^\mu \qquad (3.16)$$

is the Fourier transform of Φ_μ (see Equation (2.3)), and

$$R_k = 0(\theta^{2\mu}) \quad \text{k even} \qquad (3.17)$$

$$= 0(\theta^{2\mu+1}) \quad \text{k odd}$$

it follows that $K(\theta,h)$(appearing in the definition of π) has the form

$$K(\theta,h) = \frac{1}{h} \frac{g_1}{g_0}$$

$$= \frac{1}{h} (\theta + 0(\theta^{2\mu+1}))$$

and hence the Galerkin method using B-splines of order μ is accurate to order 2μ.

The stability of equations (3.5) will be determined by the stability of the solution to the linearized system (3.13). For all $\theta \in [-\pi,\pi]$, we require that the eigenvalues of the matrix in (3.13) have non-positive real parts; for if any eigenvalue attains a positive real part for some θ, we can expect the corresponding spatial Fourier component in the full system (3.5) to grow unstable, at least in a linear sense, until perhaps the nonlinear terms become effective in limiting the amplitude.

The θ- and h-dependence of π arises from K alone (Equation (3.10)), which plays the role of a wavenumber. This can be seen by writing the original equations (2.1) in terms of its Fourier modes; making use of the periodicity on [0,1], we define

$$\hat{X}_n(t) = \int_0^1 X(\xi,t)e^{-i2\pi n\xi}d\xi, \text{ etc,} \qquad (3.18)$$

and find

$$\dot{\hat{X}}_n = \sigma(\hat{Y}_n - \hat{X}_n)$$

$$\dot{\hat{Y}}_n = (r + 2\pi n\epsilon)\hat{X}_n - (1 - ie)\hat{Y}_n - \sum_m \hat{X}_m \hat{Z}_{n-m} \qquad (3.19)$$

$$\dot{\hat{Z}}_n = -b\hat{Z}_n + \frac{1}{2}\sum_m [\hat{X}_m(\hat{Y}_{m-n})^* + (\hat{X}_{m-n})^*\hat{Y}_m]$$

Therefore, comparing (3.8) and the second equation of (3.19), we see that wavenumber $K_n \equiv 2\pi n$ associated with a component of the partial differential equations (2.1) corresponds to K defined in Equation (3.10) associated with the finite difference scheme (3.5). However, whereas K_n can vary over an infinite range, in the numerical scheme K is limited to values such that $|K| \leq K_{max}$, where

$$K_{max} = \frac{1}{h} \max_{0 < \tau < \pi} \left| \frac{g_1(\theta)}{g_0(\theta)} \right| \qquad (3.20)$$

The value of K_{max} is thus dependent on the mesh refinement h as well as on the order of spline basis used, though this latter dependence is only a slowly varying function of the order μ. For quadratic splines,

$$\frac{g_1}{g_0} = \frac{10(2 + \cos^2\theta)}{(16 + 13\cos\theta + \cos^2\theta)} \sin\theta \qquad (3.21)$$

and

$$K_{max} = 2\pi(0.38h^{-1}) = 2\pi(0.38N) \qquad (3.22)$$

We conclude that the numerical representation of the solution to the partial differential system (2.1), given by solving Equations (3.5), can converge to the true solution only if the eigenvalues of the matrix in (3.13) have negative real parts for $K > K_{max}$. The characteristic equation for the system (3.13) is

$$(\lambda + b).\det(\pi - \lambda I).\det(\pi^* - \lambda I) = 0 \qquad (3.23)$$

Now, the root $\lambda = -b < 0$ represents a stable solution, so we must consider the other two factors. The solution of $\det(\pi - \lambda I) = 0$ is given by

$$\lambda = \frac{-(\sigma+1) + ie \pm \sqrt{(\sigma+1)^2 - e^2 - 4\sigma(1 - r_1 + \bar{Z} - \epsilon K) + 2_i(\sigma+1)(2\omega - e)}}{2} \qquad (3.24)$$

where (Fowler et al., 1982)

$$\omega = \frac{\sigma(e + r_2)}{\sigma + 1} \qquad (3.25)$$

represents the frequency of the limit cycle in complex Lorenz equations. If we write the discriminant in (3.24) as $\Delta = \alpha + i\beta$, then $\sqrt{\Delta} = \gamma + i\delta$, where

$$\gamma = \pm \frac{1}{\sqrt{2}}[\alpha + \sqrt{\alpha^2+\beta^2}]^{\frac{1}{2}}; \qquad \delta = \frac{\beta}{2\gamma} \qquad (3.26)$$

Hence, both eigenvalues in (3.24) have non-positive real parts if

$$[\alpha + \sqrt{\alpha^2+\beta^2}]^{\frac{1}{2}}/\sqrt{2} \leq \sigma + 1 \qquad (3.27)$$

By defining the critical Rayleigh number (Fowler et al., 1982) as

$$r_K = 1 + \frac{(e+r_2)(e-\sigma r_2)}{(\sigma+1)^2} \qquad (3.28)$$

$$= 1 + \omega(e-\omega)/\sigma \qquad (3.29)$$

we can reduce (3.27) to the simple form

$$\epsilon K \leq \bar{Z} - (r_1 - r_K) \qquad (3.30)$$

Now, when considering the roots of $\det(\pi^* - \lambda I) = 0$, the analysis proceeds in the same way but with e replaced by $(-e)$ and r_2 by $(-r_2)$, which reverses the sign on ω but leaves r_{1c} in (3.28) or (3.29) the same. Hence, we again arrive at (3.30), which thus represents the stability criterion for either the numerical scheme (3.5) or (by replacing K by K_n) the original system (2.1).

We note that, in either the analytic or numerical case, the replacement of r_1 by $r_1' = r_1 + \epsilon K$ (or $r_1' = r_1 + \epsilon K_n$) allows the stability criterion to be expressed as

$$r_1' - r_{1c} \leq \bar{Z} \qquad (3.31)$$

This criterion is the most complete that can be given by a linear analysis, and does not predict the value of $\bar{Z}$ which is determined by the nonlinear coupling in the third equation of (2.1) or (3.5). Intuitively, we may expect that, since in the numerical scheme, K is limited to values less than K_{max}, given by Equation (3.22) for quadratic splines, the solution may go unstable if N is chosen sufficiently large. The numerical results reported in the following section confirm this expectation.

4. NUMERICAL RESULTS-REAL LORENZ EQUATIONS

In (ABM), the real Lorenz equations were examined for small numbers of mesh points $N \leq 20$. In all cases studied, the solution was found to be asymptotic to the form

$$X(\xi,t) = R(t) \exp\{i(2\pi m\xi + const.)\}$$
$$Z = Z(t) \qquad\qquad\qquad (4.1)$$
$$\Phi_t = 0$$

with $m \leq 6$; R a (real) solution to the ordinary Lorenz equations with modified Rayleigh number $r_1' = r_1 + 2\pi m\epsilon$, and Φ_t the phase velocity of X. The form (4.1) is easily shown to be an exact solution to the Lorenz equations (2.1) for the real case ($e = r_2 = 0$). Furthermore, for the real Lorenz equations with spatial variation, there exists an integral relationship (ABM, Equation (16))

$$\int_0^1 |X(\xi,t)|^2 \Phi_t d\xi = const. \exp\{-(\sigma+1)t\} \qquad (4.2)$$

Since the integral tends to zero as $t \to \infty$, either $\Phi_t \to 0$ uniformly across the mesh, or it remains finite with changes of sign. In the cases examined in (ABM), only the former alternative, corresponding to the spatially homogeneous solutions (4.1), was found to occur. Moreover, the single remaining Fourier mode m was found to depend on the mesh refinement, with m increasing as N did. In all these cases, the stability criterion (3.30) was satisfied.

Further integrations with larger values of N reveal that this is not the only case possible. Tables 1 and 2 show that, for certain parameter values and N sufficiently large, spatially homogeneous solutions cannot be obtained, and in such cases the "energy" is not all absorbed by only one Fourier mode m, but is distributed over several modes. Furthermore, Φ_t does not decay to zero. In all these cases, the stability criterion (3.30) was found not to be satisfied.

A general tendency, for both stable and unstable solutions, was that the (spatial) average value of Z increased with the number of meshpoints N: it was only for large enough N that this growth was insufficient to counteract the destabilizing effect of the higher wave numbers appearing in the more refined mesh. Moreover, the value of m in (4.1) for the stable solution in each case corresponded to a wavelength less than or equal to the maximum wavenumber that could be supported by the mesh, which, from Equation (3.22), is

$$m_{max} = [0.38N] \qquad (4.3)$$

where [x] = integral part of x. In the case of the unstable solutions, the maximum wavenumber appeared always to contain a finite amount of energy, which was shared, in a time-dependent fashion, with other lower wavenumber modes.

This property, of the solution depending on the refinement of the mesh, is a direct consequence of the nature of the Equations (2.1), and clearly is an undesirable feature of any model for physical behaviour. It is due, in part, to there being no mechanism inherent in (2.1) for damping higher wave modes, as exists in the more usual equations of physics which contain second-order spatial derivatives and time derivatives of lower or equal order (for example, the TDGL Equation).

5. A TWO-MODE INTERACTION MODEL FOR THE LORENZ EQUATIONS.

The interaction of modes in the spatially varying Lorenz equations, which leads to either stable or unstable solutions, can be investigated in greater detail by examining a simplified model in which only two Fourier modes are present. There are two points to note about the Fourier-component form of the Lorenz equations: from Equations (3.19) we see that

(i) If only one X- and Y- mode is present, say for $n = K$, then all Z-modes with $n \neq 0$ decay to zero like $\exp(-bt)$. Furthermore, the real part of the Rayleigh number r_1 is shifted in value to $r_1' = r_1 + 2\pi K \varepsilon$. Hence, the system behaves, as $t \to \infty$, like an ordinary Lorenz system with Rayleigh number r_1'. (This 'smoothing' property for Z was the basis for choosing $Z = \bar{Z}$ in Equation (3.8), when discussing the linear stability of the numerical scheme).

(ii) If we choose only a finite set of modes of X and Y, say $K_1, K_2, \ldots, K_n$, then Equations (3.19) can be written in closed form (no truncation) provided we include all modes $\hat{Z}_j$ of Z, with $j = |K_r - K_s|$, $r, s = 1, \ldots, n$, including $j = 0$. The negative values of j are already accounted for, since equation (3.19) for $(d\hat{Z}_j/dt)$ is the complex conjugate of the equations for $(d\hat{Z}_{-j}/dt)$, on account of Z being real.

The simplest finite set of modes to consider, according to (ii), is the two-component model $n = 2$. For this case, we note that, by (i), the Rayleigh number can be re-defined as, for example,

$$r_1' = r_1 + 2\pi K_1 \varepsilon \qquad (5.1)$$

and the Rayleigh number for the second mode can be expressed in terms of it as

$$r_1'' = r_1' + 2\pi M\epsilon \qquad (5.2)$$

where,

$$M = K_2 - K_1 \qquad (5.3)$$

so that the 2-mode interactions are parametrized by the single quantity $2\pi M\epsilon$. Therefore, on writing x_j for x_{Kj}, etc., and Z_{12} for Z_M, the Fourier mode equations become, from (3.19):

$$\dot{\hat{X}}_1 = \sigma(\hat{Y}_1 - \hat{X}_1)$$

$$\dot{\hat{Y}}_1 = -(\hat{X}_2 \hat{Z}_{12}^* + \hat{X}_1 \hat{Z}_0) + (r_1' + ir_2)\hat{X}_1 - (1-ie)\hat{Y}_1$$

$$\dot{\hat{X}}_2 = \sigma(\hat{Y}_2 - \hat{X}_2)$$

$$\dot{\hat{Y}}_2 = -(\hat{X}_1 \hat{Z}_{12} + \hat{X}_2 \hat{Z}_0) + (r_1'' + ir_2)\hat{X}_2 - (1-ie)\hat{Y}_2 \qquad (5.4)$$

$$\dot{\hat{Z}}_{12} = -b\hat{Z}_{12} + \frac{1}{2}(\hat{X}_2 \hat{Y}_1^* + \hat{X}_1^* \hat{Y}_2)$$

$$\dot{\hat{Z}}_0 = -b\hat{Z}_0 + \frac{1}{2}(\hat{X}_1 \hat{Y}_1^* + \hat{X}_1^* \hat{Y}_1) + \frac{1}{2}(\hat{X}_2 \hat{Y}_2^* + \hat{X}_2^* \hat{Y}_2)$$

These constitute an untruncated set of 11 real equations, which are more amenable to numerical experimentation than the more general Galerkin system (3.5).

6. NUMERICAL RESULTS FOR THE TWO-MODE MODEL

Equations (5.4) were integrated several times, for both real and complex Lorenz equations. In the real case, the values of r_1' and M (see Equations (5.2) and (5.3)) were chosen so that r_1' and r_1'' were situated on both sides of the critical value

$$r_{crit} = \sigma(\sigma+b+3)/(\sigma-b-1) \qquad (6.1)$$

at which the classical (real) Lorenz equations show a transition to aperiodic behaviour. In all cases, M was chosen to be positive, so that mode 2 belongs to a higher wavenumber than mode 1. For the complex case, r_1' and r_1'' were chosen so that, if the modes were uncoupled, each would show a stable limit cycle behaviour as in the ordinary (complex) Lorenz equations (Fowler et al., 1982).

(i) <u>Real case</u>: The parameter values were fixed at $b = 1$, $\sigma = 3$, $r_2 = e = 0$, $\varepsilon = 1/2\pi$. The results are displayed in the $(\hat{X},\hat{Y})$ plane in Figures 1-5. When both modes are subcritical $(r_1', r_1'' < r_{crit})$, as in Fig. 1, or both supercritical (Fig. 5), it is evident that a decoupling takes place with the smaller wave number decaying to zero and the other evolving according to the ordinary Lorenz equations. (On the other hand, with $M = 1$ in the supercritical case, the modes were strongly coupled with neither showing any tendency to decay.) When mode 1 is slightly subcritical and mode 2 critical (Fig. 2), the coupling is sustained and the individual orbits are complicated. Evidence of the strong interaction are the numerous "epicycles" in each of the orbits. When one mode is subcritical and the other supercritical (Fig. 3), then a more unexpected behaviour occurs: mode 1 at first decays to a very small orbit about the origin, then suddenly revives; in the meantime, mode 2 circles the two fixed points at $\pm(\sqrt{21}, \sqrt{21})$ with frequent excursions between them. In this case, $M = 2$ and the coupling is evidently weaker than the case of Fig. 2, where $M = 1$. In figure 4, where the two modes still lie on either side of criticality but now $M = 3$, decoupling occurs on a short timescale, with mode 1 decaying to zero and mode 2 remaining finite.

(ii) <u>Complex case</u>: Parameter values chosen were $b = 0.8$, $\sigma = 2$, $r_2 = 0.1$, $e = 0.3$ and $\varepsilon = 1/2\pi$. The general property of strong coupling for small M, and weaker coupling as M is increased, is again evident. The orbits of the two modes are displayed in Figures 6 and 7 in the complex $\hat{X}$ plane. If each mode were to evolve according to the ordinary Lorenz equations, then they would tend to limit cycles with radii $|\hat{X}_1| = 7.0$ and $|\hat{X}_2| = 7.071$ (for $M = 1$) or 7.212 (for $M = 3$). The interaction between the modes significantly distorts the orbits from the circular orbits expected in the decoupled case. The limit cycle period for modes 1 and 2 in Figs. 6 and 7 is 23.56: this is, approximately, the time required to circle the origin. Superposed on this is a shorter-period oscilation that is due to the coupling, and its period seems to depend on M.

As in the real case, the lower wavenumber mode loses "energy" to the higher mode, and the large the value of M the quicker this transfer takes place.

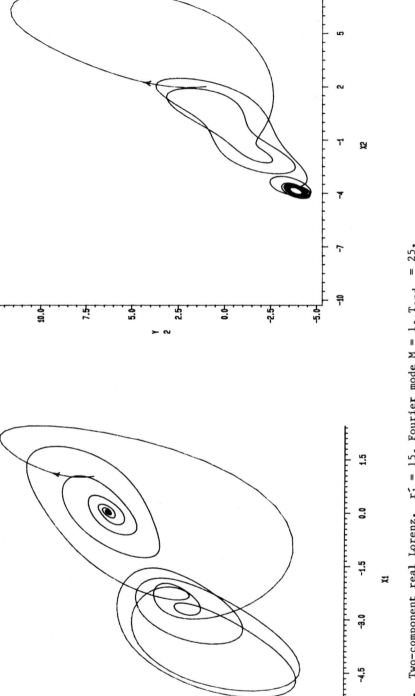

FIGURE 1. Two-component real Lorenz. $r_1' = 15$, Fourier mode $M = 1$, $T_{end} = 25$.

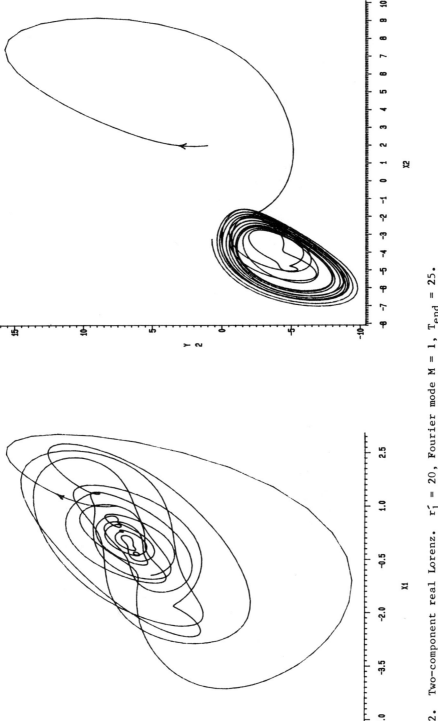

FIGURE 2. Two-component real Lorenz. $r_1' = 20$, Fourier mode $M = 1$, $T_{end} = 25$.

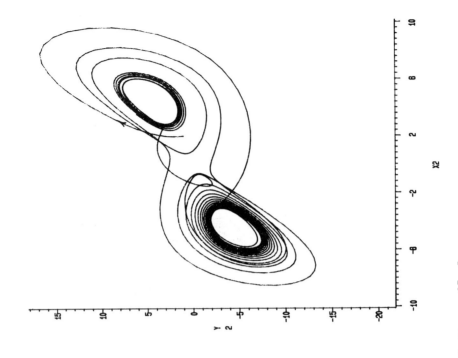

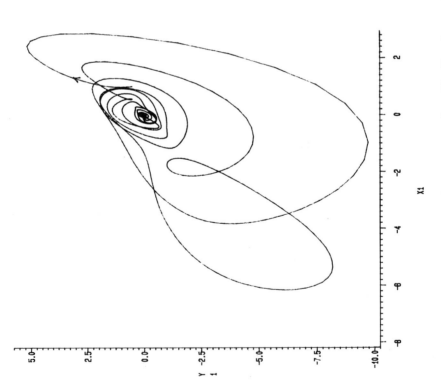

FIGURE 3. Two-component real Lorenz. $r_1' = 20$, Fourier mode $M = 2$, $T_{end} = 37.5$.

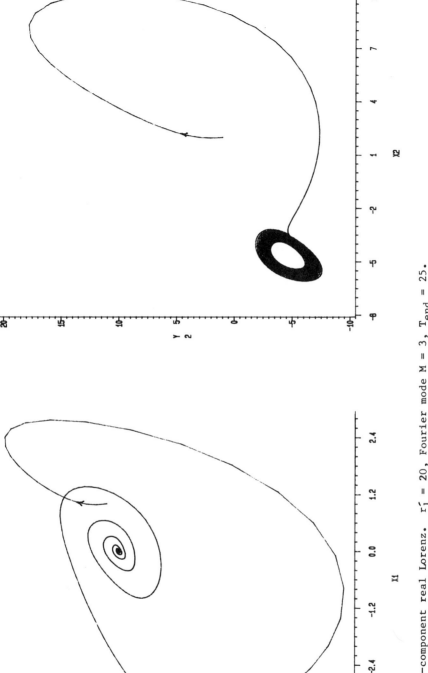

FIGURE 4. Two-component real Lorenz. $r_1' = 20$, Fourier mode $M = 3$, $T_{end} = 25$.

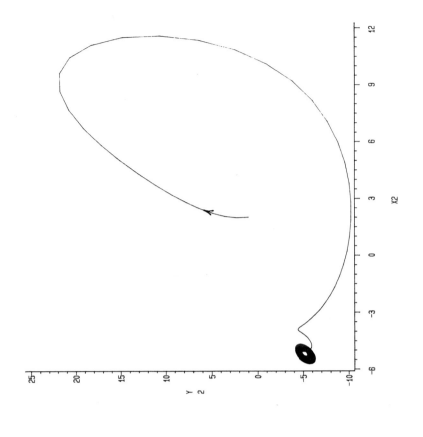

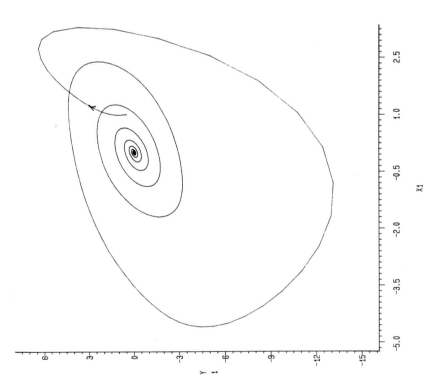

FIGURE 5. Two-component real Lorenz. $r_1' = 25$, Fourier mode $M = 3$, $T_{end} = 25$.

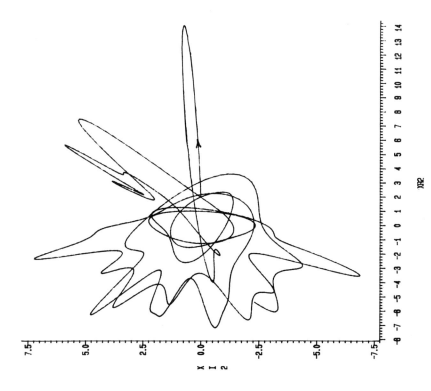

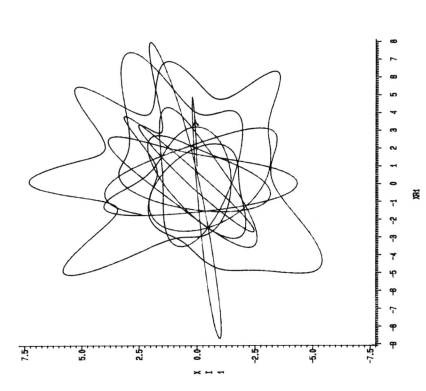

FIGURE 6. Two-component real Lorenz. $r_1' = 50$, Fourier mode $M = 1$, $T_{end} = 25$.

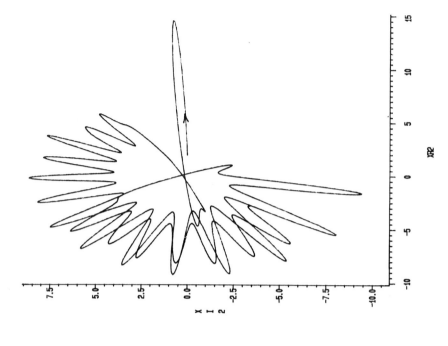

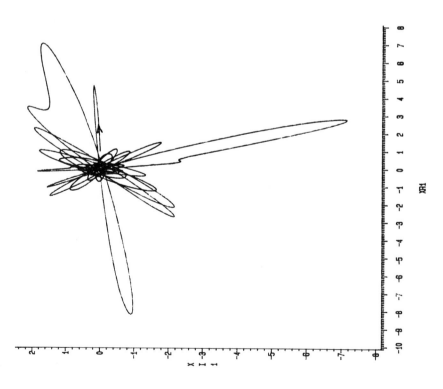

FIGURE 7. Two-component real Lorenz. $r_1' = 50$, Fourier mode M = 3, $T_{end} = 25$.

7. CONCLUSIONS

The phenomenon of the migration of energy from lower towards higher wavemodes, which appeared in the full Galerkin scheme (3.5), has been corroborated by the simple two-mode model discussed in Section 6. Also, this model shows that, when energy is shared between modes (as occurs, for example in the unstable solution in the Galerkin scheme), then this tends to occur between neighbouring modes: this is to some extent apparent from Tables 1 and 2.

It seems likely that the phenomenon arises as a consequence of the form of the equations in which the relative orders of time and space derivatives are unusual. Equations for fluid motion in which viscous effects are included normally contain second-order space derivatives which result in the damping of high wave number modes; the very good physical reason is of course the greater effectiveness of viscous dissipation in damping smaller scales of motion. The equation studied here, on the other

TABLE 1

Real Lorenz equations with spatial variation. Parameters: $\sigma = 1.44$, $b = 0.15$, $\varepsilon = 1.0$. Unstable modes are denoted by ´U´, and an asterisk denotes a dominant mode if more than one is present. The numbers m in column 3 denote wavenumbers $2\pi m$ belonging to X. Initial conditions were: $X(\xi,0) = \cos(2\pi\xi)\exp(i4\pi\xi)$; $Y(\xi,0) = (1+2.75i)X(\xi,0)$; $Z(\xi,0) = |X(\xi,0)|^2$

r_1	N	Fourier Modes	Stability
15	10	3	
	20	7	
	30	11	
	40	14,15,16	U
50	20	7	

TABLE 2

Real Lorenz equations with spatial variation. Parameters: $\sigma = 3$, $b = 1$, $\varepsilon = 0.159155$. The same notation as in Table 1 is used, with same initial conditions.

r_1	N	Fourier modes	Stability
15	15	1,3*,5	
	30	11	
	40	(several)	U
	45	(several)	U

hand, lack second-order space derivative and we might expect, in the crudest sense, to see growth of modes roughly varying with k. Nor is it sensible, in the context of the problem, to introduce an artificial viscosity in the form of a high spatial derivative since there is no diffusion mechanism for the amplitude modulation which is well modelled in this way.

In conclusion, then it appears that the absence of damping mechanisms for high wavenumber modes, and indeed the fact that the degree of instability grows with wavenumber (leading to a migration of energy from lower to higher modes), means that the Lorenz equations including spatial variation are not a viable model for weakly nonlinear phenomena. Since the equations are derived from multiple-scales expansion techniques, there is a lower and an upper bound on the wavenumbers, beyond which the Lorenz equations cease to be a valid approximation. Furthermore, the migration of energy out of the range of valid wavenumbers means that the equations can apply only for limited intervals of time, though this constraint is already inherent in the multiple- scales approximation procedure.

REFERENCES

Alexander, M. E., J. Brindley and I. M. Moroz: 1982, Phys. Lett. 87A, 240-244.

Alexander, M. E. and J. Ll. Morris: 1979, Journ. Comput. Phys. 30, 428-451.

Brindley, J. and I. M. Moroz: 1980, Phys. Lett. 77A, 441-444.

Fowler, A. C., J. D. Gibbon and M. J. McGuinness: 1982, Physica 4D, 139-163.

Gibbon, J. D. and M. J. McGuinness: 1980, Phys. Lett. 77A, 295-299.

Lange, C. G. and A. C. Newell: 1974, SIAM Journ. Appl. Math. 27, 441.

Lorenz, E. N.: 1963, Journ. Atmosph. Sci. 20, 130-141.

Moroz, I. M.: 1981, Ph.D. Thesis, University of Leeds, England.

Moroz, I. M. and J. Brindley: 1983, Studies in Appl. Math. 70, 21-62

Pedlosky, J.: 1979, "Geophysical Fluid Dynamics", Springer-Verlag.

Pedlosky, J.: 1981, Journ. Fluid Mech. 102, 169-209.

Schoenberg, I. J.: 1946, Quart Appl. Math. 4, 45-99.

Thomée, V.: 1973, in : Lecture Notes in Mathematics, Vol. 333; Springer-Verlag.

Watson, G. A.: 1979, "Approximation Theory and Numerical Methods", Wiley; Chapter 8.

10

Some Results on Singular Delay-Differential Equations

Shui - Nee Chow
Mathematics Department
Michigan State University
East Lansing, Michigan

David Green, Jr.
Science and Mathematics Department
GMI Engineering & Management Institute
Flint, Michigan

This paper gives some results on one dimensional delay-differential equations. Numerical studies are given to show how small changes in independent parameters give rise to chaotic behavior in the solution.

In this paper, we will report some recent theoretical and numerical results on one-dimensional delay-differential equations of the form

$$\varepsilon \dot{x}(t) = -x(t) + f(x(t-1), \mu) \tag{1}$$

or

$$x(t) = \frac{1}{2\varepsilon} \int_{t-1-\varepsilon}^{t-1+\varepsilon} f(x(s), \mu)ds \tag{2}$$

where $\varepsilon > 0$ is a small parameter, μ is an independent parameter, and f is a continuous nonlinear function.

Equations such as (1) or (2) have been encountered recently in many physical applications. For example, the equation that describes an optically bistable device is

$$\varepsilon \dot{x}(t) = -x(t) + \mu[1 - \sin x(t-1)].$$

In [10] Ikeda showed numerically and it was confirmed experimentally by
Gibbs, Hopf, Kaplan and Shoemaker [5] that instability or chaotic behavior
for small ε and certain values of μ occurs. In the study of evolutionary
biology, Wazewska-Czyzewska and Lasota [11] showed that the equation that
describes the production of red blood cells is

$$\varepsilon \dot{x}(t) = -x(t) + \mu x(t-1)^8 e^{-x(t-1)}.$$

Other examples can also be found in Glass and Mackey [3] and Hoppensteadt
[9]. In the second reference, examples described by equation (2) could
also be found.

Observe that when $\varepsilon = 0$, we formally obtain from equation (1) and (2)
a relation

$$x(t) = f(x(t-1), \mu).$$

This relation could be considered as a difference equation

$$x_{n+1} = f(x_n, \mu). \tag{3}$$

While (1) or (2) defines a flow in some infinite dimensional phase space,
(3) can be thought of as a discrete dynamical system, or mapping, in a one
dimensional space. The properties of system (3), readily obtained, should
have a bearing on those of the more complicated system (1); on the other
hand, (1) or (2) may exhibit new phenomena not found in the simpler system
(3).

More precisely, if $x(\theta) = \phi(\theta)$, $-1 \leq \theta \leq 0$, then for $\varepsilon > 0$ (1)
determines a unique solution $x(t)$ for $t \geq 0$. This is found simply by
integrating the equation

$$\varepsilon \dot{x}(t) = -x(t) + g(t), \quad 0 \leq t \leq 1$$
$$x(0) = \phi(0),$$

where $g(t) = f(x(t-1), \mu)$, to obtain $x(t)$ for $0 \leq t \leq 1$. Successive
integrations determine $x(t)$ for all $t \geq 0$. As in Hale [8], let $C = C[-1,0]$ be the Banach space of all continuous functions on the interval
$[-1,0]$ with the usual sup norm. If $y(\cdot)$ is a continuous function on some

interval, then let y_t denote the element in C defined by $y_t(\theta) = y(t+\theta)$, $-1 \leq \theta \leq 0$. Thus, given $x_0 = \phi \in C$ we uniquely obtain $x(t)$ for $t \geq 0$ and therefore $x_t \in C$ for $t \geq 0$. This may be formalized by introducing the nonlinear evolution map

$$T(t,\varepsilon) : C \to C \qquad t \geq 0$$

given by $T(t,\varepsilon)\phi = x_t$. As (1) is autonomous, we have that $T(t,\varepsilon)T(s,\varepsilon) = T(t + s, \varepsilon)$, for all t, $s \geq 0$.

Consider now the time one map $S(\varepsilon) = T(1,\varepsilon)$. We have for any $\phi \in C$ and $-1 \leq \theta \leq 0$,

$$s(\varepsilon)\phi(\theta) = \phi(0)e^{\frac{-\theta + 1}{\varepsilon}} + \frac{1}{\varepsilon}\int_{-1}^{\theta} e^{\frac{-\theta - s}{\varepsilon}} f(\phi(s), \mu)ds.$$

Thus, for any $\delta > 0$ we have uniformly on $[-1 + \delta, 0]$,

$$S(\varepsilon)\phi(\theta) \to f(\phi(\theta),\mu), \text{ as } \varepsilon \to 0$$

In a gross sense, therefore, $S(\varepsilon)$ is close to the mapping f. In terms of the solution $x(t)$, we have for any $n \geq 1$ and $\delta > 0$ fixed,

$$x(n + \theta) \to f^{(n)}(x(\theta), \mu), \text{ as } \varepsilon \to 0$$

uniformly for $\theta \in [-1 + \delta, 0]$, where $f^{(n)} = fo\cdots of$ is the nth iterate of f. In this way, (3) may be considered as a singular limit of (1).

A similar argument may be made for equations (2) and (3). In the following, we will describe some results relating these equations. We note that the actual relation between the discrete system (3) and the continuous system (1) or (2) is still not clearly understood.

We first consider equation (2). Assume that $f \in C^2$ and μ is fixed (so that we supress the μ variable in f). Let f satisfy

$$\left\{ \begin{array}{ll} f(-x) = f(x) \\ f(x) < -x, & 0<x<1 \\ f'(0) < -1 \\ f'(x) < 0, & x \in \mathbb{R} \\ f''(x) > 0, & x>0 \end{array} \right. \qquad (4)$$

It is not difficult to see that for the discrete system (3), there exists
a stable period-2 orbit {1,-1}. It is natural to make the conjecture that
there exists a stable periodic orbit of equation (2) with least period
which is approximately equal to 2. Recently, Chow, Diekmann and Mallet-
Paret [1] showed that this is indeed the case. In fact, one has the
following theorem.

Theorem 1 Let $0<\epsilon<1/3$ and $f\epsilon C^2$ satisfy condition (4). Then for every
$0<\epsilon<1/3$, there exists a periodic solution ϕ_ϵ of (2) with period exactly 2.
Moreover, ϕ_ϵ satisfies the following symmetry conditions.

$$\phi(t+1) = -\phi(t)$$
$$-\phi(t) = \phi(t) \hspace{3cm} (5)$$
$$\phi(t) \geq 0, \quad t\epsilon[0,1]$$

and is the unique periodic solution of (2) within the class of periodic
functions satisfying (5). Furthermore, ϕ_ϵ is monotone
in ϵ, i.e., $\phi_{\epsilon_1}(t) \leq \phi_{\epsilon_2}(t)$ provided $t\epsilon[0,1]$ and $\epsilon_1 \geq \epsilon_2$, and
each ϕ_ϵ is uniformly asymptotically stable.

In Theorem 1, the bound 1/3 is only used in showing the stability of
ϕ_ϵ. We can relax the condition to the case $0<\epsilon<1/2$ for the existence and
uniqueness of the ϕ_ϵ's. Numerical studies indicate that the stability
property holds for $1/3<\epsilon<1/2$. It is also possible to construct examples
in which $\epsilon=1/2$ and there are no periodic solutions for (2) with f
satisfying (4). The proof of stability of ϕ_ϵ is nontrivial.

We remark that Theorem 1 is not a typical result relating the
discrete system (3) to its continuous counterparts, equation (1) or (2).
For example, numerical studies indicate that for certain nonlinear
functions f, (3) has a stable period-2 orbit which is a global attractor
(except for a finite number of points), upon iteration, while there are no
periodic solutions of (1) with least period approximately 2, for any $\epsilon>0$
which is arbitrarily small. This, of course, is rather surprising.

However, the situation is rather nice when the discrete system (3)
has an attracting fixed point.

Suppose for some μ, a is an exponentially stable fixed point of f
under iteration; that is,

$$f(a,\mu) = a, \hspace{2cm} |f_x(a,\mu)| < 1. \hspace{2cm} (6)$$

By considering the characteristic equation of the linearized differential-delay equation we have the following.

Theorem 2. Suppose (6) holds. Then for any $\varepsilon > 0$, $x(t) \equiv a$ is a uniformly asymptotically stable solution of (1) or (2).

Suppose the fixed point a of f loses its stability as μ passes through μ_1. We expect the stability of the constant solution $x(t) \equiv a$ will also change as μ passes through μ_1. In fact, we have the following.

Theorem 3. Suppose that $f(a,\mu) = a$ and $f_x(a,\mu) = -1 -\alpha(\mu-\mu_1) + 0(|\mu-\mu_1|^2)$ for μ near μ_1 where $\alpha>0$ is a constant. Then the constant solution $x(t) \equiv a$ of (1) or (2) loses its stability along a curve Γ

$$\alpha(\mu-\mu_1) = \frac{\pi^2}{2} \varepsilon^2 + 0(\varepsilon^3), \ (\mu,\varepsilon) \text{ near } (\mu_1,0), \qquad (7)$$

on which a pair of purely imaginary eigenvalues near $\pm i\pi$ of the characteristic equation of the linearization of (1) or (2) at $x(t) \equiv a$ occurs. Indeed, $x(t) \equiv a$ is exponentially stable for parameters to the left of Γ, and unstable to the right, in a neighborhood of $(\mu,\varepsilon) = (\mu_1, 0)$.

Details may be found in Chow and Mallet-Paret [2] and Chow, Diekmann and Mallet-Paret [1].

It can be shown by the methods in [3] and [7] that a Hopf bifurcation occurs along the curve Γ (7). Furthermore, numerical studies indicate that the bifurcations are often supercritical for classes of equations arising in applications. The bifurcating stable periodic solutions may be approximated by $a + c \sin(\pi t)$, where $c > 0$ is the amplitude, for $(\mu-\mu_1,\varepsilon)$ small and near the curve (4).

In Figures 1, 2 and 3, we illustrate the above results by considering the equation

$$\varepsilon \dot{x}(t) = -x(t) + \mu[1-\sin x] \qquad (8)$$

It is not difficult to see that at $\mu = 1.5$ the discrete system

$$x_n = \mu[1-\sin x] \qquad (9)$$

has a stable period-2 orbit. For $\varepsilon = 0.5$ and $\mu = 1.5$, (8) has a stable constant solution (see Figure 1), i.e., (1.5, 0.5) is above the curve Γ.

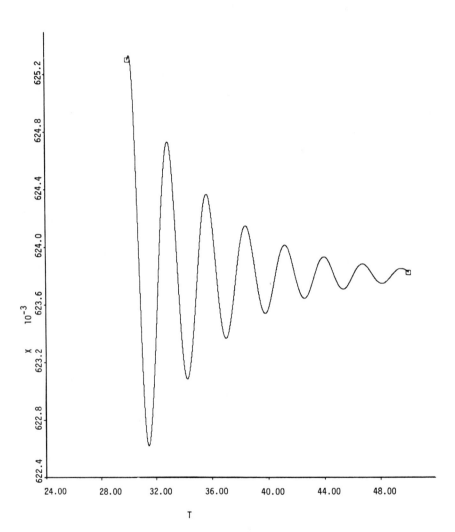

FIG. 1. $\mu = 1.5$, $\varepsilon = .5$, initial function $\phi = 1$.

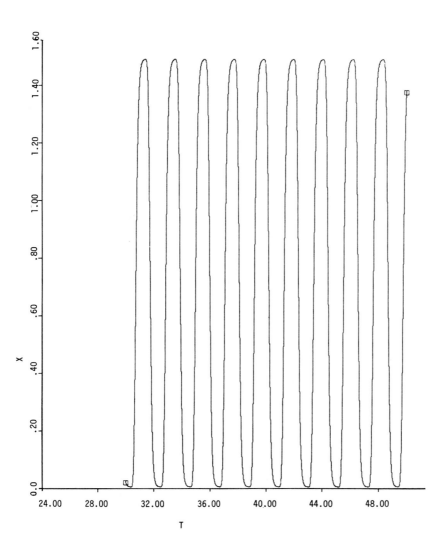

FIG. 2.   μ = 1.5,   ε = .07125, initial function ϕ = 1.

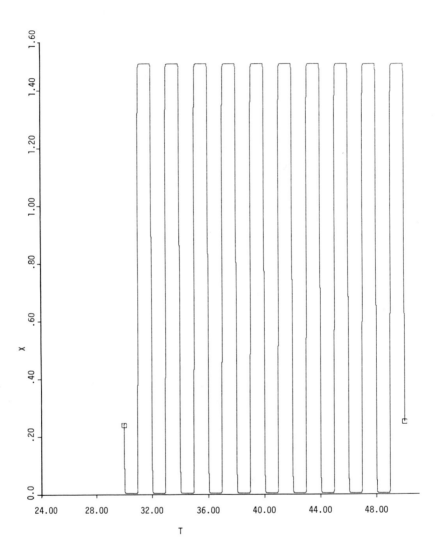

FIG. 3. $\mu = 1.5$, $\varepsilon = .01$, initial function $\phi = 1$.

On the other hand, (1.5, 0.07125) and (1.5, 0.01) are below the curve Γ and for these values of μ and ε we have a sine wave and a square wave respectively. (see Figures 2 and 3).

At μ ≅ 1.96544335, the period-2 orbit bifurcates into a stable period-4 orbit. In Figures 4 and 5, we observe the existence of a stable periodic orbit whose least period is approximately 4. It is stable because we used different initial data in Figures 4 and 5 and they all seem to converge to the same periodic orbit. When ε is increased to 0.5, the square wave periodic orbit in Figure 5 slowly goes into a sine wave whose least period is close to 2. This is illustrated in Figure 6.

In Figures 7, 8 and 9, we illustrate the existence of period-8 and period-16 orbits. It seems that system (8) undergoes a Feigenbaum Cascade of bifurcations [4] just as the discrete system (9) does. Note that the period-16 orbit is also stable.

However, numerical computer studies show that this relation does not occur between (2) and (3) with

$$f(x,\mu) = -\mu x + x^3.$$

Figures 10 and 11 illustrate the chaotic behavior observed as μ is increased. Note that in Figures 12, 13, and 14 the solution is no longer chaotic. As ε is increased the solution becomes sinusoidal instead. We also note that in Figures 4-9, periodic solutions exhibit certain abrupt changes. These behaviors can be interpreted by a transition layer equation [3].

To better understand the periodic behavior in these systems, we consider equation (1) in more detail. We assume $f(a_0, \mu) = a_1$ and $f(a_1,\mu) = a_0$ for some $a_0 < a_1$. Furthermore, we assume that the period-2 orbit $\{a_0,a_1\}$ is globally stable. In Chow and Mallet-Paret [3] a transition layer equation is introduced to describe the transition of a periodic solution of (1) from a_0 to a_1 or a_1 to a_0. To obtain the equation, we introduce the scaled time variable τ:

$$\varepsilon\tau = -t$$

As in [3] we assume that $\phi_\varepsilon(t)$ is a periodic solution of (1) with least period $2 + 2k\varepsilon$, where k is some unknown constant. Since (1) is

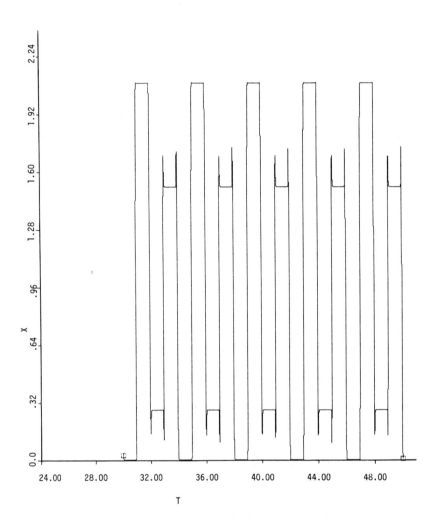

FIG. 4. $\mu = 2.1$, $\varepsilon = .01$, initial function $\phi = 1$.

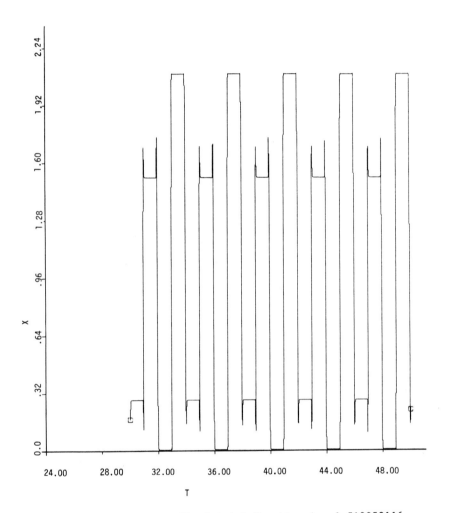

FIG. 5.   μ = 2.1, ε = .01, initial function φ = 1.518053116.

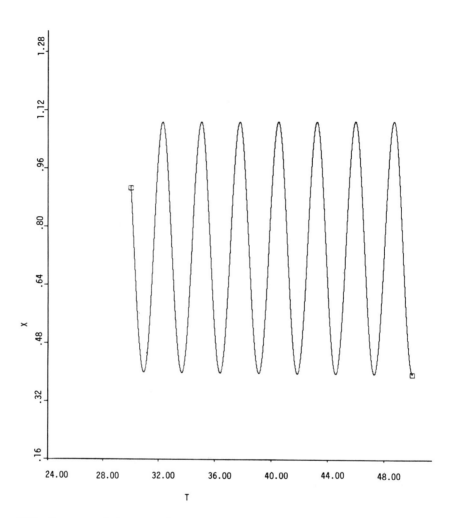

FIG. 6. $\mu = 2.1$, $\varepsilon = .5$, initial function $\phi = 1$.

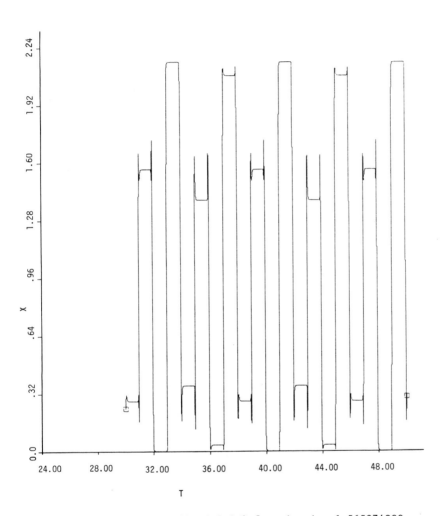

FIG. 7.   μ = 2.16, ε = .01, initial function φ = 1.565374889.

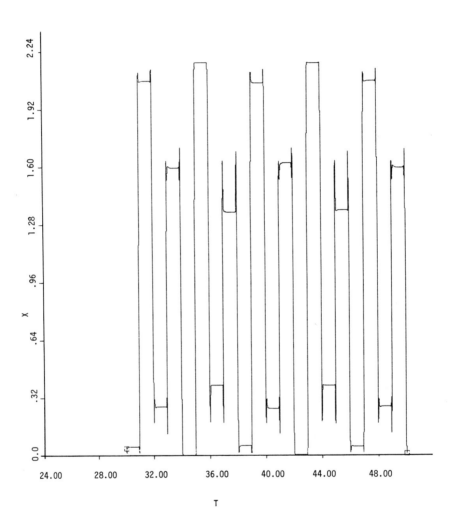

FIG. 8. $\mu = 2.181$, $\varepsilon = .01$, initial function $\phi = 2.0757745$.

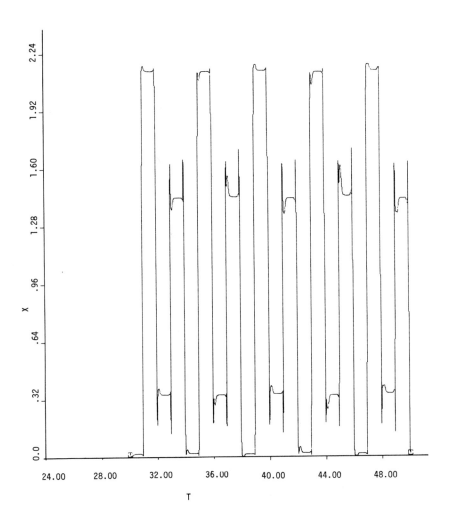

FIG. **9.**   μ = 2.181, ε = .01, initial function φ = 1.

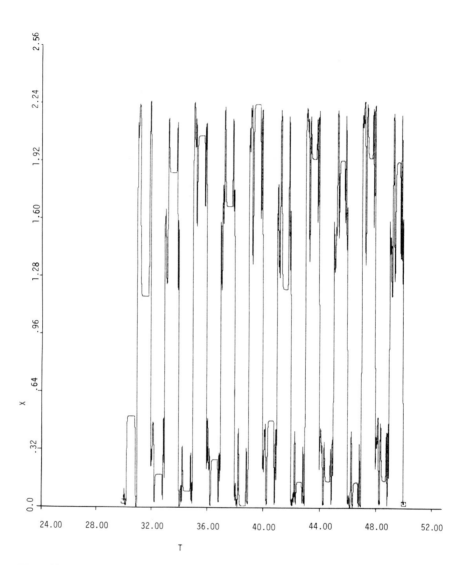

FIG. 10. $\mu = 2.26$, $\varepsilon = .01$, initial function $\phi = 1$.

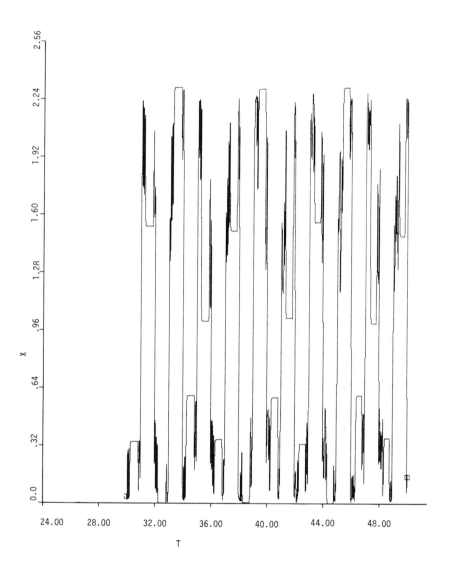

FIG. 11.   μ = 2.31, ε = .01, initial function φ = 1.

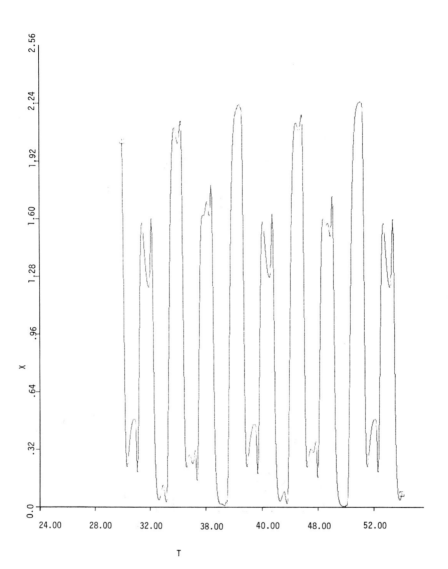

FIG. 12. $\mu = 2.26$, $\epsilon = .1$, initial function $\phi = 1$.

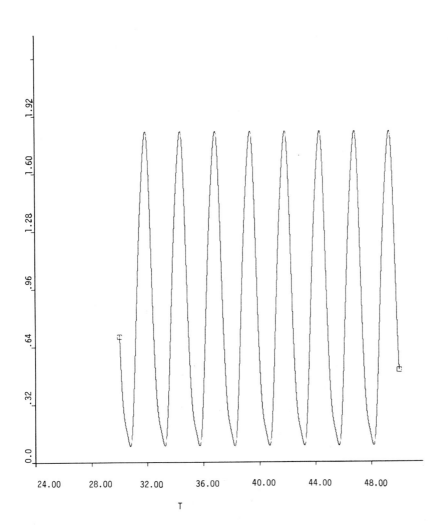

FIG. 13. μ 2.26, ε = .3, initial function φ = 1.

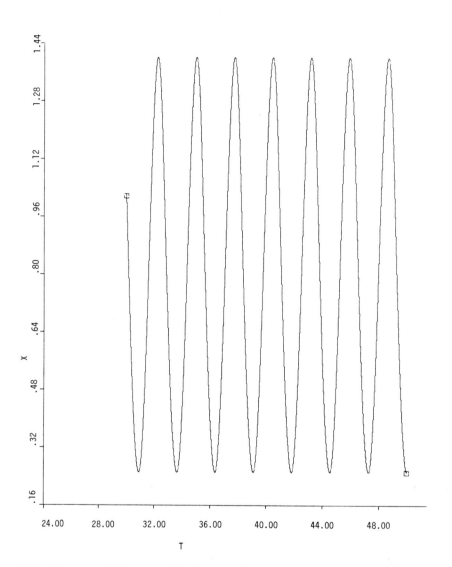

FIG. 14. $\mu = 2.31$, $\varepsilon = .5$, initial function $\phi = 1$.

autonomous, we may assume that $\phi_\varepsilon(0) = 0$. Let

$$y_\varepsilon(\tau) = \phi_\varepsilon(-\varepsilon\tau)$$

and

$$z_\varepsilon(\tau) = \phi_\varepsilon(1 + k\varepsilon - \varepsilon\tau)$$

It can be shown that there exists a subsequence $\varepsilon_n \to 0$ and y and z such that

$$y_{\varepsilon_n} \to y \text{ and } z_{\varepsilon_n} \to z \text{ as } \varepsilon_n \to 0$$

uniformly on every compact interval. Moreover, y and z satisfy the transition layer equation

$$\frac{d}{d\tau} y(\tau) = y(\tau) - f(z(\tau-k), \mu)$$

$$\tag{10}$$

$$\frac{d}{d\tau} y(\tau) = z(\tau) - f(y(\tau-k), \mu)$$

with boundary conditions

$$(y(\tau), z(\tau)) \to (a_1, a_0), \tau \to -\infty$$

$$(y(\tau), z(\tau)) \to (a_0, a_1), \tau \to +\infty$$

$$\tag{11}$$

Note that the problem of solving (10) and (11) consists of finding the unknown k for which there exists a heteroclinic orbit of (10). In [1] and [3], existence and continuation of such heteroclinic orbits are considered. In some sense, this is a generalization of Melnikov's method in two-dimensional systems and the global bifurcation of periodic orbits.

By using the transition layer equation (10) and the discrete system (3), it is possible in certain cases to obtain a matching of these solutions, producing a periodic solution of least period approximately 2 of equation (1). The question of stability is still open. However, numerical studies strongly indicate that a theorem such as Theorem 1 is possible for equation (1).

REFERENCES

1.   S. N. Chow, O. Diekmann and J. Mallet-Paret, in preparation.
2.   S. N. Chow and J. Mallet-Paret, J. Diff. Eq., 26 (1977), 112-159.
3.   S. N. Chow and J. Mallet-Paret, Singularly perturbed delay-differential equations, Physica D., to appear.

4.   M. J. Feigenbaum, Quantitative universality for a class of nonlinear
      transformations, J. Stat. Phys. 19 (1978), 25-52; 21 669-706, 1979.
5.   H. M. Gibbs, F. A. Hopf, D.. Kaplan and R.L. Shoemaker, Observation
      of chaos in optical bistability. Phys. Rev. Lett., 46 (1981), 474-
      477.
6.   L. Glass and M. Mackey, Oscillation and chaos in physiological
      control systems. Science, 197 (1977), 287-289.
7.   D. Green, Self-Oscillations for epidemic models, Bull. Math. Biol.,
      38 (1978), 91-111.
8.   J. Hale, Theory of functional differential equations, Springer-
      Verlag, (1977).
9.   F. C. Hoppensteadt, Mathematical Theories of Population:
      Demographics, Genetics and Epidemics. S.I.A.M. publ., Phila., (1976).
10.   K. Ikeda, Multiple-valued stationary state and its instability of the
      transmitted light by a ring cavity systems. Opt. Commun. 30, 257-
      574(1979).
11.   A. Losota, Ergodic problems in biology. Soc. Math. France, Aster.,
      50 (1977), 239-250.

11

Feigenbaum Functional Equations as Dynamical Systems

P. Fischer

Department of Mathematics and Statistics
University of Guelph
Guelph, Ontario, Canada

Dynamical systems generated by even and unimodal solutions of the Feigenbaum functional equations are studied. With the aid of a construction continuum many even and unimodal solutions of the equation is exhibited for all λ, $0 < \lambda < 1$. It is pointed out that the nature and the structure of the periodic points of the above solutions (which are only differentiable almost everywhere) is the same as of the ´smooth´ solutions.

1. One of the simplest example for dynamical systems can be exhibited with the aid of iterations of maps of an interval into itself. There is a vast literature on this subject.

Recently, many papers dealt with Feigenbaum´s conjectures concerning a special class one-parameter of family of smooth unimodal transformations. We are not describing in details Feigenbaum conjectures, since by now, there are some excellent accounts of this topic [2], [3], [5]. We are considering only the case $\varepsilon = 1$. Our work is a direct continuation of [4], where some of the results discussed in this note were presented without proof.

In his work about the Feigenbaum functional equation Lanford [5] is considering the equation of the form

(1) $f(f(\lambda x)) + \lambda f(x) = 0, \; \lambda = -f(1)$

where the solution f is unimodal, even and fairly smooth on $[-1,1]$ and such that

(2) $f(0) = 1; \; f(1) < 0; \; f(\lambda) \geq \lambda, \; f''(0) < 0.$

At the end of his paper Lanford remarks that he finds it to be quite unsatisfactory that we don't know why a solution exists.

A natural way to proceed to answer that question seems to be to analyse (1) from functional equation's point of view. That method was initiated in [4]. In the present paper we continue that method and some further properties of the solutions will be obtained. We also present a construction of a family of unimodal solutions of (1) for each λ, $0 < \lambda < 1$. These solutions are differentiable only almost everywhere, nevertheless their invariant Cantor sets have the same structures as the ones of the smooth solutions.

2. We shall say that a mapping f of $[-1,1]$ into $[-1,1]$ is unimodal (see [2]) if
α) $f \varepsilon \; C[-1,1]$
β) $f(0) = 1$
γ)   f is strictly increasing on $[-1,0]$ and strictly decreasing on $[0,1]$.
We shall use the following notations:
$f_2 = fof, \; \ldots, \; f_{n+1} = fof_n$, for $n = 1,2,\ldots$
A point x is said to be a periodic point of period m (or a period m point) if $f_m(x) = x$, but $f(x) \neq x,\ldots,f_{m-1}(x) \neq x$, and $m \geq 2$.
        The following theorems had been proven in [4].
Theorem 1   Let f be an even unimodal solution of (1), with $-1 < f(1) < 0$. Then f has a unique fixed point x_1, where $x_1 > 0$, and

$$\lambda < x_1 < f(\lambda)$$

Theorem 2   If f is an even unimodal solution of (1), with $-1 < f(1) < 0$, and if f does not have a period 2 point in the interval $(\lambda, f(\lambda))$, then x_1

is an unstable fixed point of **f**. Furthermore, if x ε $(\lambda, f(\lambda))$, x≠x_1, then
the orbit of x leaves this interval in finitely many steps, and there is
no periodic point of any period in $(\lambda, f(\lambda))$.

3. The following theorem gives a regularity condition, which ensure that
f will not have a period 2 point in $(\lambda, f(\lambda))$. Note, that we do not use
the concept of the Schwarzian derivatives.

Theorem 3 If f is an even unimodal solution of (1) with $-1 < f(1) < 0$
and if f is twice differentiable on $[-1,1]$, $f''(0) < 0$ and $f''(x) \leq 0$ on
$[0,1]$, then f has no period 2 point in $(\lambda, f(\lambda))$.

Proof Differentiating (1) yields that

(3) $f'(f(\lambda x))f'(\lambda x) = -f'(x);$

from which one obtains that

(4) $f'(f(\lambda))f'(\lambda) = -f'(1).$

Using the fact that $f''(0) < 0$ we can conclude that

$f'(1) = -\frac{1}{\lambda}$. Hence (4) becomes

(5) $f'(f(\lambda))f'(\lambda) = \frac{1}{\lambda}$.

Combining (5) with the fact that f' is nonincreasing on $[0,1]$, we can
conclude that $f' \leq -1$ on $[\lambda, f(\lambda)]$. Next, we shall show that $f' < -1$ on
$[\lambda, f(\lambda)]$. If it were not so, then we would have $f'(\lambda) = -1$ and

$f'(f(\lambda)) = -\frac{1}{\lambda}$. Hence $f' = -\frac{1}{\lambda}$ for xε$[f(\lambda),1]$.

Using the fact that $f(\lambda x)$ε$[f(\lambda),1]$ for xε$[0,1]$, we see that (3)
reduces to

(6) $f'(\lambda x) = \lambda f'(x)$

for xε$[0,1]$. Then, by induction, we have that $f' = -1$ on $[\lambda f(\lambda),\lambda]$, $f' = -\lambda$ on $[\lambda^2 f(\lambda),\lambda^2]$,...,$f' = -\lambda^n$ on $[\lambda^{n+1} f(\lambda), \lambda^{n+1}]$,...

Now
$$\lim_{n \to \infty} \frac{f'(0)-f'(\lambda^{n+1})}{-\lambda^{n+1}} = -\frac{1}{\lambda}$$

but
$$\lim_{n \to \infty} \frac{f'(0)-f'(\lambda^{n+1}f(\lambda))}{-\lambda^{n+1}f(\lambda)} = -\frac{1}{\lambda f(\lambda)}, \quad \text{which}$$

would imply that f''(0) does not exist, in contradiction to our hypothesis.

Now, if it were exist points α and β so that $\lambda < \alpha < \beta < f(\lambda)$ with $f(\alpha) = \beta$, $f(\beta) = \alpha$, then by the mean – value theorem we would get

$$\left| f(\alpha)-f(\beta) \right| = \beta-\alpha = \left| f'(\xi) \right| (\beta-\alpha),$$

which is impossible.

4. The attractor set of the solutions of (1) exhibits a large variety of different structures depending exclusively on the behavior of f on $(\lambda, f(\lambda))$. There are solutions having countably many stable periodic orbits and there are solutions having continuum many periodic orbits. One can show the existence of an invariant Cantor set of an arbitrary even and unimodal solution of (1) with $-1 < \lambda < 0$, which does not have a period 2 point in $(\lambda, f(\lambda))$. The construction described in [3] remains also valid in that case.

We shall state without proof, the only additional result which is needed in the proof of this generalized result.

Theorem 4   If f is an even unimodal solution of (1) with $-1 < \lambda < 0$, then we have that

(7) $\left| f_{2^{n+1}+2^n}(0) \right| > \left| f_{2^n+1}(0) \right|$,

and

(8) $\min_{1 \le i \le 2^n} \left| f_i(0) \right| = \left| f_{2^n}(0) \right|$

for $n = 0, 1, \cdots$.

Note that the statement of (7) is equivalent to Theorem 1 in the case when n = 0.

The next theorem describes the nature of the periodic points of a unimodal solution of (1). It is a partial generalization of Proposition 8.4 of [3], since the only regularity condition is imposed is that there is no periodic 2 point in $(\lambda, f(\lambda))$.

Theorem 5 If f is an even unimodal solution of (1) with $-1 < f(1) < 0$, and if f does not have a periodic point of period 2 in $(\lambda, f(\lambda))$, then

(i) f has exactly one periodic orbit of 2^m period (with $m = 1,2...$).

(ii) Every orbit of f which does not eventually fall exactly on one of the repelling periodic orbits enumerated in (i) converges to the invariant Cantor set.

Proof. We shall sketch the proof of (i) for the periodic orbit 2. The functional equation (1) shows that $-\lambda x_1$ is a periodic point of order 2, and it can be shown using only (1), that if x_1 is an unstable fixed point of f, then $- \lambda x_1$ is an unstable period 2 point. Now we shall show that $\{-\lambda x_1, f(-\lambda x_1)\}$ is the only period 2 orbit of f.

Indeed, if it were exist another period 2 orbit then one member of this orbit, say μ, must be in the interval $(-\lambda, \lambda^2)$. By letting $\lambda x = u$ in (1) we obtain

$$(9) \qquad f_2(u) + \lambda f(\frac{u}{\lambda}) = 0 \qquad \text{for} \quad -\lambda \leq u \leq \lambda,$$

which shows that $- \frac{\mu}{\lambda}$ is a fixed point of f. But this is impossible, since f has an unique fixed point.

This argument can be iterated using a remark of [1] and noting that $(-\lambda)^n x_1$ is a period 2^n point. The proof of (ii) is essentially the same as presented in [3].

5. It is easy to see that (1) does not have an even unimodal solution for $\lambda = 1$. In the light of that fact it is surprising that the cardinality of the unimodal even solutions of (1) for each fixed λ, $0 < \lambda < 1$, is the continuum. We shall show that by an appropriate construction. This method is in itself interesting, because that way one might be able to obtain a larger class of solutions.

The idea is that we start with a function defined at first only on $[0, f(\lambda)]$, and then to extend that function so that, the extended function to be an even, unimodal solution of (1) on [-1,1]. In our example, we are starting with a piecewise linear function, satisfying the conclusions of Theorem 1, and we can extend this function uniquely with the aid of (1). Hence, it seems quite plausible, if one has a ´good´ starting function, a larger class of solutions can be obtained.

Let λ be fixed, $0 < \lambda < 1$. We define f to be $1 - \mu_0 x$ on $[0,\lambda]$, where $\mu_0 > 0$ and such that $f(\lambda) = 1 - \mu_0\lambda > \lambda$, and we define f to be linear on $[\lambda,f(\lambda)]$ such that $f(\lambda)$ as before and $f(f(\lambda)) = \lambda^2$.

Now, since f has to satisfy (1), it follows that we must have

(10) $f(1-\mu_0\lambda x) = -\lambda f(x)$ for $0 \leq x \leq 1$.

Let $T(x) = 1-\mu_0\lambda x$. With the aid of (10), we can define f on the intereval $[T(1-\mu_0\lambda), T(0)]$. Repeating the same steps we can define f on the interval $[T_2(0), T_2(1-\mu_0\lambda)]$ and so one, and

finally we define $f(\frac{1}{1+\mu_0\lambda}) = 0$. It is quite interesting to see, that in this way we have obtained a function defined over the whole interval $[0,1]$, which can be extended to $[-1,1]$ so that the extended function be an even unimodal solution of (1).

We note that it is easy to show that the solutions constructed in that way do not have periodic point of order 2 in the interval $(\lambda,f(\lambda))$. Hence the nature and the structure of the periodic points does not separate the solutions of (1) according to their smoothness.

References

1.   M. Campanino, M. Epstein, D. Ruelle, On Feigenbaum´s functional equation $gog(\lambda x) + \lambda g(x) = 0$, Topology, Vol. 21. No. 2., 125-129, 1982.
2.   P. Collet and J-P. Ekmann, Iterated maps on the interval as dynamical systems; Birkhauser Boston, 1980.
3.   P. Collet, J-P. Ekmann, and O.E. Lanford III, Universal Properties of Maps on an Interval; Commun. Math. Phys. 76, 211-254 (1980).
4.   P. Fischer, Feigenbaum functional equation and periodic points, to appear.
5.   O. E. Lanford III, Smooth transformations of intervals, Séminaire Bourbaki; vol. 1980/81 no. 563, Lecture Notes in Mathematics, no. 901. Springer-Verlag, 1981.
6.   O. E. Lanford III, A computer-assisted proof of the Feigenbaum conjectures. Bulletin of the A.M.S., Volume 6, Number 3, 427-434, 1982.

12

The Chaos of Dynamical Systems

Morris W. Hirsch

Department of Mathematics
University of California
Berkeley, California

In this article I first discuss the general concepts of chaotic and
nonchaotic behavior, stability, and genericity for dynamical systems.
Then the special class of monotone systems is described, with some
examples. Three theorems are stated which show that the dynamics of
monotone systems cannot be very chaotic.

    A dictionary definition of chaos is "a disordered state or
collection; a confused mixture". This is an accurate description of
dynamical systems theory today -- or of any other lively field of
research.

    Having defined chaos, let me define the other key words in the title.
In the style of Euclid, who said a point is that which has no part, I
define a system to be anything with more than one part. A dynamical
system is one which changes with time. What changes is the state, that
is, the relationship between the parts of the system. (One could, and
should, consider systems whose state spaces change with time; but this
seems very difficult to handle mathematically.)

To understand a dynamical system means to know how its states vary through time -- at least to describe their variation, at best to predict it.

The oldest dynamical system under study is the cosmos, whose parts are the heavenly bodies. Many mathematical models of this system have been proposed, with radically different dynamics ranging from the epicycles of Ptolemy to the geodesics of Einstein.

Mathematical dynamical systems theory had its inception with Newton. Before Newton more or less ad hoc geometrical methods were used to describe the dynamics of the cosmos. Newton started from a physical theory -- universal gravitation and laws of force -- and derived differential equations which determined the dynamics. Poincaré and then G.D. Birkhoff initiated a more abstract and topological study of differential equations. Ever since Birkhoff, dynamical systems theory in the mathematical sense has meant the study of the long run behavior of solutions to differential equations in which one variable is thought of as time. Of course the subject has undergone considerable internal development, and today it includes fairly remote generalizations and abstractions from the original differential equation model.

An old and all-important idea in both applied and abstract dynamics is that of stability. A widely accepted principle, called the Stability Dogma in Abraham-Marsden (1967), is that since measurements and numerical calculations are never exact, only those features of a system which persist under perturbations are considered to have physical (or biological, chemical, etc.) significance. Such features are called "stable" or "robust." Many kinds of stability have been studied; in particular a feature may be stable under perturbations of the state within a fixed system, or alternatively it may be stable under perturbations of the whole system within some space of systems.

Closely related to stability is the concept of generic behavior. Since it is a hopeless task to classify all possible dynamical systems in any broad field of interest, a plausible strategy is to try to describe the behavior of the "typical" system. This attractive idea, initiated by Pontryagin and Andronov and emphasized by Smale and his school, is fraught not only with mathematical difficulties, but also with practical and philosophical ones. For one thing, it has been firmly established (among others, by Smale) that generic behavior, under any reasonable definition, need not be stable. Furthermore it is often far from clear which definition of generic is reasonable. For example one often takes the definition of a generic to mean a Baire set (intersection of a countable family of open dense sets) in some suitable topology on the set of systems or states in question. If the state space is the space of real numbers then this definition makes the set of irrationals a generic set. The difficulty now arises that only rational numbers can be measured, or programmed on a computer. Another reasonable definition of generic set is one whose complement has measure zero (for some appropriate measure). But this is not the same as the first definition, and in fact a set can be generic under this definition while its complement is generic under the first definition! This problem is discussed in Kaplan-Yorke (1979), who refer to Arnold (1965).

An interesting example of chaos -- in several senses -- is provided by the celebrated Lorenz System:

$$dx/dt = -10x + 10y$$

$$dy/dt = 28x - y - xz$$

$$dz/dt = -\frac{8z}{3} + xy$$

This is an extreme simplification of a system arising in hydrodynamics. By computer simulation Lorenz (1963) found that trajectories seem to wander back and forth between two particular stationary states, in a random, unpredictable way. Trajectories which start out very close together eventually diverge, with no relationship between long run behaviors.

But this type of chaotic behavior has not been proved. As far as I am aware, practically nothing has been proved about this particular system. Guckenheimer (1976), Williams (1979) proved that there do indeed exist many systems which exhibit this kind of dynamics, in a rigorous sense; but it has not been proved that Lorenz's systems is one of them. It is of no particular importance to answer this question; but the lack of an answer is a sharp challenge to dynamicists, and considering all the attention paid to this system, it is something of a scandal.

The Lorenz system is an example of (unverified) chaotic dynamics; most trajectories do not tend to stationary or periodic orbits, and this feature is persistent under small perturbations. Such systems abound in models of hydrodynamics, mechanics, and many biological systems. On the other hand experience (and some theorems) show that many interesting systems can be expected to be nonchaotic: most chemical reactions go to completion; most ecological systems do not oscillate unpredictably; the solar system behaves fairly regularly. In purely mathematical systems we expect heat equations to have convergent solutions, and similarly for a single hyperbolic conservation law, a single reaction diffusion equation, or a gradient vector field.

A major challenge to mathematicians is to determine which dynamical systems are chaotic and which are not. Ideally one should be able to tell from the form of the differential equations. The Lorenz system illustrates how difficult this can be.

There is an interesting class of dynamical systems which are guaranteed to be nonchaotic, and which can usually be easily recognized. These are what I call <u>monotone systems.</u> In order to describe them I need some definitions.

For simplicity I consider dynamical systems whose state space is an open set W in some (real) Banach space X. The dynamic is given by a <u>flow</u> ϕ: a collection of continuous maps $\{\phi_t : W_t \to W\}_{t \geq 0}$ defined in open subsets of W, having the following properties:

$W_0 = W$ and $\phi_0 =$ identity map of W;

$\phi_t^{-1}(W_s) = W_{s+t}$ and $\phi_s \phi_t = \phi_{s+t}$;

the map $(t,x) \to \phi_t(x)$ is continuous and defined on an open set in $[0,\infty) \times W$ which contains $\{0\} \times W$.

Now suppose X is (partially) ordered by a cone

$$X_+ = \{x \; \varepsilon \; X: \; x \geq 0\}.$$

<u>I assume that the interior of of X_+ is nonempty.</u> When X is a function space this is a rather delicate condition. Write

$$x \geq y \text{ if } x - y \; \varepsilon \; X_+$$

$$x > y \text{ if } x \geq y \text{ and } x \neq y$$

$$x \gg y \text{ if } x - y \; \varepsilon \text{ int } X_+.$$

For example Euclidean n-space R^n is ordered by saying $x \geq y$ if $x_i \geq y_i$ for all i. Then $x \gg y$ means $x_i > y_i$ for all i. If X is the Banach space of C^1 functions on a compact Riemannian manifold which vanish on the boundary, with the ordering $f \geq g$ meaning $f(x) \geq g(x)$ everywhere, then $f \in$ int X_+ precisely when $f(x) > 0$ on the interior of the manifold and the outward normal derivative of f is strictly negative at the boundary.

A map f between ordered Banach spaces is called <u>monotone</u> in case $x \geq y$ implies $f(x) \geq f(y)$, and <u>strongly monotone</u> if $x > y$ implies $f(x) \gg f(y)$.

Now let ϕ denote a flow in an open set W of an ordered Banach space. I call ϕ

<u>monotone</u> if ϕ_t is monotone for each $t \geq 0$, and

and

<u>strongly montone</u> if ϕ_t is strongly monotone for each $t > 0$.

An example of a monotone flow is one generated by a C^1 vector field F in R^n having the property that $\partial F_i / \partial x_j \geq 0$ for $i \neq j$; this follows from a celebrated theorem of Kamke (1932) (see also Coppel (1965), Walter (1970)). The systems used by Lajmanovich and Yorke (1976), to model the gonorrhea epidemic have this property, which I call <u>cooperative</u>. If the Jacobian matrices $[\partial F_i / \partial x_j]$ are also irreducible then the flow can be proved to be strongly monotone.

For another example consider a semilinear parabolic partial differential equation of the form

$$\partial u / \partial t = \Delta u + F(x,u,\nabla u); \quad t > 0, \ x \in \bar{\Omega}$$

with standard boundary conditions (e.g. Dirichlet or Neumann), in a smooth bounded domain $\Omega \subset R^n$. Under mild restrictions on f there is a flow ϕ in the space $C_B^1(\bar{\Omega})$ of C^1 functions on $\bar{\Omega}$ satisfying the boundary conditions, such that if $u(x,t)$ is the solution with initial data $u(x,0) = v \in C_B^1(\bar{\Omega})$, then $u(x,t) = \phi_t(v)(x)$. Standard maximum principles imply that ϕ is strongly monotone. This flow has the additional property of being <u>order-compact</u>. This means that for $t > 0$, ϕ_t maps any order interval [u,v] into a precompact set. (See Mora (1982), Smoller (1983), Henry (1981).)

Now let ϕ be a flow in W. An <u>attractor</u> K is a nonempty compact set having a neighborhood N in W such that every trajectory starting in N has compact closure in W and has all its limit points in K. Chaotic dynamical systems are sometimes said to have "strange attractors" or even "strange

strange attractors" (Guckenheimer (1976)). This is a somewhat vague concept, but usually means: K is neither a single stationary point (= equilibrium) or a single periodic orbit, and periodic orbits are dense in K, and some one orbit is dense in K.

Theorem 1. An attractor for a monotone flow contains a stationary point. It cannot contain a dense orbit. Periodic nonstationary orbits can not be dense in K.

In other words there are no strange attractors in a monotone flow.

From now on let ϕ be strongly monotone flow which is order-compact, in which every orbit has compact closure in W.

Theorem 2. Let K be an attractor. Then K contains a stable equilibrium. If the set of equilibria in K is finite then some equilibrium is asymptotically stable. If K contains only one equilibrium p, then p attracts every trajectory attracted to K.

A trajectory is quasiconvergent if it is attracted to the set of equilibria. It is convergent if it converges to an equilibrium. When the equilibrium set is countable these two concepts are the same.

The following result is another sense in which strongly monotone flows are nonchaotic.

Theorem 3. Let A ⊂ W be a totally ordered topological arc. Let B ⊂ A be the subset of points of A that are not quasi-convergent. Then B is countable and discrete in the relative topology.

This theorem can be interpreted as saying that if a point is chosen at random then with probability one its trajectory is quasiconvergent.

For more details, applications, examples, and related work, see the articles listed in the bibliography by Amann, Conlon, Hirsch, Yorke et al., Othmer, Selgrade, and Smale. For earlier uses of monotonicity in differential equation see the survey by Amann (1976). Much useful material on partial differential equations as dynamical systems can be found in the books of Henry, Walker, Smoller, Martin, Marsden and McCracken, and Hassard, Kazarinoff and Wan.

BIBLIOGRAPHY

Abraham, R. and Marsden J. 1967. Foundations of Mechanics, (First edition). Addison Wesley, New York.

Amann, H. 1976. Fixed point equations and nonlinear eigenvalue problems in ordered Banach spaces, SIAM Rev. 18, 620-709.

Amann, H. 1978. Perodic solutions of semilinear parabolic equations, in Nonlinear Analysis (ed. by L. Cesari, R. Kannan, H. Weinberger), Academic Press, New York.

Amann, H. 1976. Existence and multiplicity theorems for semi-linear elliptic boundary value problems, Math. Z. 150, 281-295.

Arnold, V. I. 1965. Small denominators, I. Translations of Amer. Math. Soc. 2nd Series 46, 213-284.

Conlon, J. G. 1980. A theorem in ordinary differential equations with an application to hyperbolic conservation laws, Adv. Math. 35, 1-18.

Coppel, W. A. 1965. Stability and Asymptotic Behavior of Differential Equations, D.C. Heath, Boston.

Crandall, M. G. and Majda, A. 1980. Monotone difference approximations for scalar conservation laws, Math. Comp. 34, 1-21.

Grossberg, S. 1978. Competition, decision and consensus, J. Math. Anal. Appl. 66, 470-493.

Guckenheimer, J. A. 1976. Strange strange attractor. In: The Hopf Bifurcation Theorem and its Applications. J.E. Marsden and M. McCracken, Eds. Springer-Verlag, New York.

Henry, D. 1981. Geometric Theory of Semilinear Parabolic Equations, Lecture Notes in Mathematics 840. Springer-Verlag, New York.

Hirsch, M. W. 1982. Systems of differential equations which are competitive or cooperative. I: Limit sets, SIAM J. Math. Anal. 13, 167-179.

Hirsch, M. W. 1982a. Differential equations and convergence almost everywhere in strangly monotone flows. In: Nonlinear Partial Differential Equations (J. Smoller, ed.), Contemporary Math. 17. Amer. Math. Soc., New York.

Hirsch, M. W. 1985. Systems of differential equations which are competitive or cooperative. II: Convergence almost everywhere. SIAM J. Math. Anal. 16, 432-439.

Hassard, B. D., Kazarinoff, N. D. and Wan, Y.-H. 1981. Theory and Application of Hopf Bifurcation, London Mathematical Society Lecture Note Series 41, Cambridge University Press, Cambridge, England.

Kamke, E. 1932. Zur theorie der systeme gewohnlicher differential-gleichungen, II, Acta Math. 58, 57-85.

Kaplan, J. and Yorke, J. 1979. The onset of chaos. In: Bifurcation Theory and its Applications, O. Gurel and O. Rossler, Eds. Annals of N.Y. Acad. Sci. Vol. 16, 400-407.

Lajmanovich, A. and Yorke, J. 1976. A deterministic model for gonorrhea in a nonhomogeneous population, Math. Biosci. 28, 221-236.

Lorenz, E. 1963. Deterministic nonperiodic flow. J. Atmos. Sci. 20, 130-141.

Martin, R. H. 1976. Nonlinear Operators and Differential Equations in Banach Spaces, John Wiley & Sons, New York.

Mora, X. 1982. Semilinear problems define semiflows on C^k spaces, Math. Dept., U. Michigan, (Preprint).

Othmer, H. G. 1976. The qualitative dynamics of a class of biochemical control circuits", J. Math. Biol. 3, 53-78.

Protter, M. H. and Weinberger, H. 1967. Maximum Principles in Differential Equations, Prentice-Hall, Englewood Cliffs, New Jersey.

Selgrade, J. 1979. Mathematical analysis of a cellular control process with positive feedback, SIAM J. Appl. Math. 36, 219-229.

Selgrade, J. 1960. Asymptotic behavior of solutions to single loop positive feedback systems, J. Diff. Eq. 38, 80-103.

Smale, S. 1976. On the differential equations of species in competition, J. Math. Biol. 3, 5-7.

Smoller, J. 1983. Shock Waves and Reaction Diffusion Equations, Springer-Verlag, New York.

Walker, J. A. 1980. Dynamical Systems and Evolution Equations, Plenum Press, New York.

Walter, W. 1970. Differential and Integral Inequalities, Springer-Verlag, New York.

Williams, R. F. 1979. Bifurcation space of the Lorenz attractor. In: Bifurcation Theory and its Applications, O. Gurel and O. Rossler, Eds. Annals of N.Y. Acad. Sci. Vol. 16, 393-399.

13

On Network Perturbations of Electrical Circuits and Singular Perturbation of Dynamical Systems

Gikō Ikegami[*]

Department of Electrical Engineering
University of Waterloo
Waterloo, Ontario, Canada

Dynamical systems of electrical networks including jumping behavior have been reduced locally to singular perturbation theory for ordinary differential equations. This paper reduces the situations globally to the geometric singular perturbation theory of N. Fenichel. As an application, the jumping behaviors of states of electrical networks are given in neighborhoods of compact subsets of regular domains.

1. Introduction

The purpose of this paper is to give the foundation of a general theory of dynamical systems arising from electrical circuits which involves the theory of relaxation oscillations [4,7].

   A theory of an electrical circuit N is specified by a vector $(i,v) \in X = C_1(N) \times C^1(N)$ of currents and voltages of branches. By Kirchhoff's laws, (i,v) is included in a linear subspace K of X. The characteristics of resistors give the restraint that the vector (i_R, v_R)

---

*Permanent address: Dept. of Mathematics, Nagoya University, Chikusa, Nagoya , Japan

on resistor branches is included in a manifold Λ.   Generically, $\sum = K \cap \Lambda$ is called a configuration manifold.

The   dynamical   system   of N is   a   vector   field   on $\sum$ defined   by $v_L = L(i_L) \cdot di_L/dt$ and $i_C = C(v_C) \cdot dv_C/dt$, where (i_L, v_L) and (i_C, v_C) are vectors on inductor branches and capacitor branches, respectively.   The solution   vector x is   not   defined   at   a   singular   point   of   the   projection $\sum \to C_1 (L) \times C^1(C)$, $(i,v) \to (i_L, v_C)$, but actually, the dynamical system corresponds   to   a   relaxation   oscillation,   as   many   states   change discontinuously at a singular point.

F. Takens [8]   reduces   the   study   to   the   consideration of   singular perturbations   of   constrained   differential   equations,   in the   case when   a suitable   real-valued   function   exists.   S.S.   Sastry   and C.A.   Desoer [6] treat   jump   behavior   related to   singular   perturbations   of   constrained differential   equations   as   models of the   jump   behavior   of   electrical circuits.

In   this   paper   we   study   the   phenomenon   by   means   of   N. Fenichel's   geometric   singular perturbation   theory   of dynamical   systems [1].   The   approach   of this   paper is   to   consider   a network   perturbation defined   in   Definition   3.2,   which can   be   considered   to   be   a   natural notion.   The main theorem appears in §3.   Using a theorem of N. Fenichel [1] we show a corollary in §4 as an application.

The   proofs   of propositions and theorems are omitted in this   paper; they will appear elsewhere.

The   following   example   is well known as   a   relaxation   oscillation [4,7].   This example illustrates the use of our method.

Example.   Let N be   the   circuit shown in Fig.1,   such that   the characteristic of resistor R is given by a function $v_R = f(i_R)$ whose graph Λ_R is shown in Fig. 2.

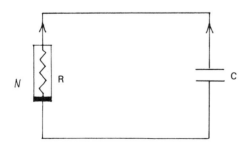

FIGURE 1

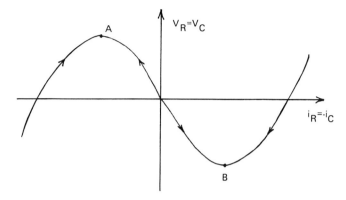

FIGURE 2

The Kirchhoff space K is determined by $v_C = v_R$ and $-i_R = i_C$.   Then the 2-plane of Fig. 2 can be considered as K, and $\sum = \Lambda_R$ is the configuration manifold. The dynamical system Y is determined

on $\sum$ by $i_C = C \cdot \dfrac{dv_C}{dt}$, where C is the capacitance of capacitor C. Then X is

not defined at the singular points A,B.   Let $\tilde{N}_\varepsilon$ be the circuit shown in Fig. 3, which is obtained by adding an inductor L of inductance εL to the circuit of Fig. 1.   This is a network perturbation of N. The Kirchhoff space $\tilde{K}$ is determined by $v_C = v_R + v_L$ and $-i_R = -i_L = i_C$. The configuration manifold $\tilde{\sum}$ is illustrated in Fig. 4. $\tilde{\sum}$ has no singular points. The dynamical system Y_ε is the vector field on $\tilde{\sum}$ determined by

$$i_C = C \cdot \frac{dv_C}{dt} \text{ and } v_L = \varepsilon L \frac{di_R}{dt}.$$

A subspace $K_0 = \{(i,v) \ \varepsilon \ \tilde{K}; \ v_L = 0\}$ of $\tilde{K}$ can be identified with K, and $\sum_0$

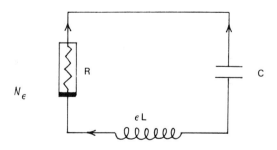

FIGURE 3

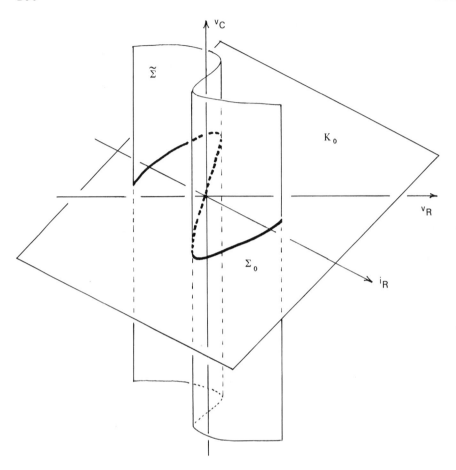

FIGURE 4

$= \sum K_0$ can be identified with $\sum$. In fact, if π is a forgotten map of i_L, π maps K_0 onto K isomorphically and $\sum_0$ onto $\sum$ diffeomorphically. We regard Y as a vector field on $\sum_0$. When $\varepsilon \downarrow 0$, the manifold does not approach any invariant manifold of Y_ε. But, considering the time descaled field $X_\varepsilon = \varepsilon Y_\varepsilon$, $\sum_0$ is an invariant manifold of X_0. In fact, each point of $\sum_0$ is an equilibrium point of X_0. $(\sum_0)_H = \sum_0/\{A,B\}$ is a normally hyperbolic invariant manifold of X_0, where A and B are the singular points of $\sum_0$. By an invariant manifold theorem of dynamical systems [2], there is a normally hyperbolic invariant manifold $\sum^\varepsilon_H$ of X_ε near $(\sum_0)_H$. If X_ε is C^r close to X_0 then $\sum^\varepsilon_H$ is C_r close to $(\sum_0)_H$. Since X_ε and Y_ε have the same orbit structure, $\sum^\varepsilon_H$ is also invariant under Y_ε. If a point $p\varepsilon\sum$ is not close to $\sum_0$ or $\sum_\varepsilon$, then the vector $Y_\varepsilon(p)$ is parallel to the i_C (or

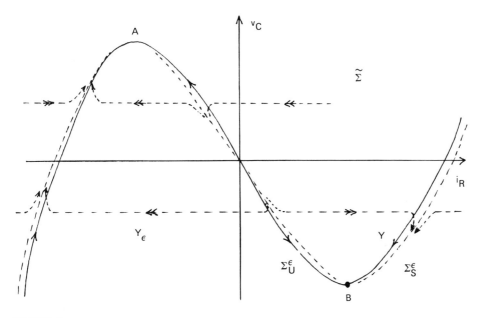

FIGURE 5

i_R)-axis, and $\|Y_\epsilon(p)\|$ is very large. If $p \in \int^\epsilon_H$, then $Y_\epsilon(p)$ is tangent to $\int^\epsilon_H$, which is close to $\int_0$. Hence, we can see, by the equation for Y_ϵ, that $Y_\epsilon(q)$, $q\epsilon\int^\epsilon_H$. Near the normally stable domain $\int^\epsilon_S \subset \int^\epsilon_H$ the trajectories of Y_ϵ jump to $\int^\epsilon_S$. A trajectory which starts at a point near the normally stable domain $\int^\epsilon_U$ jumps into $\int^\epsilon_S$ or to far away. See Fig. 5.

2. Preliminary

Let G be the oriented graph of the given electrical circuit with resistors, inductors and capacitors. Let $C_k(G)$ and $C^k(G)$ respectively be a real k-chain and a real k-cochain of G, k=0,1. A state of the circuit is specified by a current vector $i = (i_1,\ldots,i_b) \in C_1(G)$ and a voltage vector $v = (v_1,\ldots,v_b) \in C^1(G)$, where b is the number of the branches contained in G. Sometimes, we will use the notations $C_1(N)$ and $C^1(N)$ in the meaning of $C_1(G)$ and $C^1(G)$.

Kirchhoff's law restricts the possible states of N to a b-dimensional linear subspace of $C_1(G) \times C^1(G)$ called the Kirchhoff space K = Ker∂ $\times$ Im∂^*, where ∂ is the boundary operator $\partial:C_1(G) \to C_0(G)$ and ∂^* is the coboundary operator $\partial^*:C^0(G) \to C^1(G)$.

2.1 Definition. A graph $\tilde{G}$ is a __fundamental extension__ of G if $\tilde{G}$ is obtained by adding branches to G in the following manners:

(S)   Insert a finite number (may be zero) of extra branches in series with each branch of G.

(P)   Connect each pair of vertices of G by finite number (may be zero) of parallel extra branches.

Here, we set a notation for the extra branches of (S) and (P) and the vertices of boundaries as in Fig. 6.

Let T be a maximal tree of G.   Then L = G\T is the link of T.   We can take a maximal tree $\tilde{T}$ of $\tilde{G}$ such that (i) T ⊂ $\tilde{T}$,   and (ii) every extra branch inserted by (S) is contained in $\tilde{T}$.   Consequently, every extra branch γ_j added by (P) is contained in the link $\tilde{L}$ of $\tilde{T}$.   In fact, the

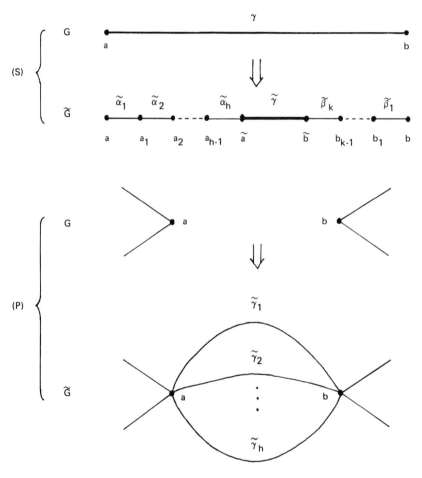

FIGURE 6

vertices a,b in the boundary of γ are contained in T, so that a,b are contained in $\tilde{T}$. Hence γ_j cannot contained in $\tilde{T}$. Therefore, we have

$$G = T \cup L, \quad \tilde{G} = \tilde{T} \cup \tilde{L} \tag{2.1}$$

$$T \subset \tilde{T}, \quad L \subset \tilde{L}.$$

Let $T' = \tilde{T} \backslash T$ and $L' = \tilde{L} \backslash L$. Then, T' and L' are the graphs consisting of the extra branches added by (S) and (P) respectively.

Let $\tilde{\partial}: C_1(\tilde{G}) \to C_0(\tilde{G})$ and $\tilde{\partial}^*: C^0(\tilde{G}) \to C^1(\tilde{G})$ be the boundary and coboundary operators, respectively. The Kirchhoff space of $\tilde{G}$ is given by $\tilde{K} = \text{Ker}\tilde{\partial} \times \text{Im}\tilde{\partial}^*$. Since $\tilde{G} = G \cup T' \cup L'$, the elements $\tilde{i} \in C_1(\tilde{G})$ and $\tilde{v} \in C^1(\tilde{G})$ are decomposed as

$$\tilde{i} = (\tilde{i}_G, \tilde{i}_{T'}, \tilde{i}_{L'}) \text{ and } \tilde{v} = (\tilde{v}_G, \tilde{v}_{T'}, \tilde{v}_{T'}). \tag{2.2}$$

Define

$$(\text{Ker}\tilde{\partial})_0 = \{\tilde{i} \in \text{Ker}\tilde{\partial}; \ \tilde{i}_{L'} = 0\},$$

$$(\text{Im}\tilde{\partial}^*)^0 = \{\tilde{v} \in \text{Im}\tilde{\partial}^*; \ \tilde{v}_{T'} = 0\} \text{ and} \tag{2.3}$$

$$K_0 = (\text{Ker}\tilde{\partial})_0 \times (\text{Im}\tilde{\partial}^*)^0.$$

Then, K_0 is a linear subspace of $\tilde{K}$.

Define the projections

$$\pi_0: C_1(\tilde{G}) \to C_1(G) \text{ and} \tag{2.4}$$

$$\pi^0: C^1(\tilde{G}) \to C^1(G)$$

by $\quad \pi_0(\tilde{i}) = \tilde{i}_G$ and $\pi^0(\tilde{v}) = \tilde{v}_G$.

2.2 Proposition. (i) π_0 maps $(\text{Ker}\tilde{\partial})_0$ isomorphically onto $\text{Ker}\partial$. (ii) π^0 maps $(\text{Im}\tilde{\partial}^*)^0$ isomorhically onto $\text{Im}\partial^*$. So that $(\pi_0 \times \pi^0)|K_0:K_0 \to K$ is an isomorphism.

Let r, ℓ and c be the number of the resistors, inductors and capacitors in the circuit N. Let R, L and C be the subgraph of G which consists of the branches that correspond to the resistors, inductors and capacitors of N, respectively. Now we have b=r+ℓ+c.

The following are the standing hypotheses of this paper:

2.3 Assumptions.

(A) The graph of a circuit is connected.

(B) A circuit is time-invariant.

(C) The resistor constitutive relations are characterized
by

$$(i_R, v_R) \in \Lambda_R,$$

where Λ_R is a r-dimensional submanifold of class C^s, $s \geq 2$, in $\mathbb{R}^{2r} \simeq C_1(R)$
$\times C^1(R)$. Λ_R is called the <u>characteristic manifold</u> of R.

In   addition,   we   will   assume   additional   hypotheses   concerning
inductors and capacitors in the next section.

We have the following natural direct sum decompositions:

$$C_1(N) = C_1(R) + C_1(L) + C_1(C) \cong \mathbb{R}^r \times \mathbb{R}^\ell \times \mathbb{R}^c$$
$$C^1(N) = C^1(R) + C^1(L) + C^1(C) \cong \mathbb{R}^r \times \mathbb{R}^\ell \times \mathbb{R}^c$$

By the hypothesis (C) above, the state $(i,v) \in C_1(N) \times C^1(N)$ of N is
contained in the submanifold.

$$\Lambda = \Lambda_R \times C_1(L \cup C) \times C^1(L \cup C) \sim \Lambda_R \times R^{2(\ell+c)}.$$

Here, the $\sim$ denotes a diffeomorphism. Let $C^r(\Lambda_R, \mathbb{R}^{2r})$ be the space of the
C^r-mappings of Λ_R into $\mathbb{R}^{2r}$ with Whitney C^r-topology, where we consider $\mathbb{R}^{2r}$
as $C_1(R) \times C^1(R)$.   Then,   generically with respect to the space
$C^r(\Lambda_R, \mathbb{R}^{2r})$;   the manifold $\Lambda = \Lambda_R \times R^{2(\ell+c)}$ and the Kirchhoff space K^b
intersect transversally in $\mathbb{R}^{2b} = C_1(N) \times C^2(N)$.   This is shown in Theorem
5 [5] in the C^1 case, but the extension to C^r is trivial.   Hence, we may
assume that $\sum = \Lambda \cap K$ is a $(\ell+c)$-dimensional C^r manifold.   We say that $\sum$
is the <u>configuration manifold</u> of the circuit N.

2.4 <u>Proposition</u>. <u>Let $\tilde{G}$ be a fundamental extension of G. Let Λ be a</u>
C^r <u>submanifold in</u> $C_1(G) \times C^1(G)$ <u>having transverse intersection with the</u>
<u>Kirchhoff space K. Put</u> $\tilde{\Lambda} \equiv \Lambda \times C_1(L' \cup T') \times C^1(L' \cup T') \subset C_1(\tilde{G}) \times C^1(\tilde{G})$,
$\tilde{\sum} \equiv \tilde{\Lambda} \cap \tilde{K}$ <u>and</u> $\sum_0 \equiv \tilde{\sum} \cap K_0$. <u>Then (i) the following intersections are</u>
<u>transverse:</u> $\tilde{\Lambda} \pitchfork K$ <u>in</u> $C_1(\tilde{G}) \times C^1(\tilde{G})$ <u>and</u> $\tilde{\sum} \pitchfork K_0$ <u>in</u> $\tilde{K}$ <u>so that</u> $\tilde{\sum}$ <u>and</u> $\sum_0$ <u>are</u>
C^r <u>manifolds,</u> <u>and</u> (ii) $\pi_0 \times \pi^0$ <u>map</u> $\sum_0$ <u>diffeomorphically onto</u> $\sum$.

3. Main Theorem

For each inductor L_j and each capacitor C_k of a circuit let i_j, v_j and
i_k, v_k be the currents and voltages of the corresponding branches.

3.1 Assumption (D):

The above defined quantities satisfy the following equations:

$$v_j = L_j(i_j) \frac{di_j}{dt} \tag{3.1}$$

$$i_k = C_k(V_k) \frac{dv_k}{dt} \tag{3.2}$$

where $L_j(i_j)$ and $C_k(v_k)$ are non-zero real-valued functions of class C^{r-1}, $r \geq 2$, which are called the <u>inductance</u> of L_j and the <u>capacitance</u> of C_k.

These are the "equations of motion" for electrical networks. Let $\pi^\prime : C_1(N) \times C^1(N) \to C_1(L) \times C^1(L)$ be the natural projection. Put

$$\pi = \pi^\prime | \textstyle\sum \tag{3.3}$$

A domain $\sum_r$ of the configuration manifold $\sum$ is <u>regular</u> if $\pi: \sum \to C_1(L) \times C^1(C)$ is regular at each point (i,v) in $\sum_r$, i.e. the differential of π at (i,v) has full rank. A point $(i,v) \in \sum$ which is not contained in any regular domain is called a <u>singular point</u>.

Since Λ_R is a C^r manifold, $r \geq 2$, by Assumption 2.3, $\pi|\sum_r : \sum_r \to C_1(L) \times C^1(L)$ is a C^r <u>diffeomorphism</u> onto the image. Then the <u>dynamical system</u> of N is defined on a regular domain $\sum_r$. This is the C^{r-1} vector field defined by

$$X(i,v) = (\pi^{-1}) * Y(\pi(i,v)), \tag{3.3}$$

where, Y is the C^{r-1} vector field on $\pi(\sum_R)$ given by (3.1) and (3.2) and $D\pi^{-1}$ is the differential of π^{-1}.

But, at a singular point (i,v) the vector $X(i,v)$ cannot be defined. This is the difficulty which the dynamics of electrical networks involves.

In Example of §1 the network N is extended by inserting an inductor of small inductance. For the purpose of generalization we introduce the following definition.

3.2 <u>Definition</u>: A circuit $\tilde{N}$ is a <u>network perturbation</u> of N if $\tilde{N}$ is obtained by adding branches to N in the following way:

(S) To every branch of N insert in series a finite number (may be zero) of extra branches of inductors with "small" inductances and resistors with "small" resistances.

(P) Connect each pair of vertices of N by a finite number (may be zero) of parallel extra branches of capacitors with "small" capacitance and resistors with "large resistances."

Here, "small" and "large" have the following meanings: Small inductance means that the function $L_j(i_j)$ in (3.1) is C^{r-1} close to the constant map $O(i_j) = 0$ in Whitney C^{r-1} topology. Small capacitance means the function $C_k(v_k)$ in (3.2) is C^{r-1} close to the constant zero map. Let R_o and R_∞ be the set of the branches of the resistances which are added to N by (S) and (P) respectively. Then "small resistances" means that the characteristic manifold $\Lambda_R \subset C_1(R_0) \times C^1(R_0)$ is C^r close to the subspace $C_1(R_0) \times \{o\} \subset C_1(R_0) \times C^1(R_0)$ in the space

$$(C^r(C_1(R_0), C_1(R_0) \times C^1(R_0))$$

of all C^r mappings, $C_1(R_0) \rightarrow C_1(R_0) \times C^1(R_0)$, with Whitney C^r topology. "Large resistances" means that $\Lambda_{R\infty} \subset C_1(R_\infty) \times C^1(R_\infty)$ is C^r close to $\{o\} \times C^1(R_\infty) \subset C_1(R_\infty) \times C^1(R_\infty)$

The graph $\tilde{G}$ of $\tilde{N}$ is a fundamental extension of the graph G of N.

Let M be a C^r manifold. Suppose $\{S_y: y \in U\}$ is a family of submanifolds of M parametrized by y in a manifold U. Let

$$S^* = \{(x,y): x \in S_y, y \in U\} \subset M \times U.$$

We say that $\{S_y: y \in N\}$ is a C^r __family of manifolds__ if S^* is a C^r submanifold of $M \times U$.

Let $\{M_\varepsilon: 0 \leq \varepsilon < \varepsilon_0\}$ and $\{N_\varepsilon: 0 \leq \varepsilon < \varepsilon_0\}$ be C^r families of manifolds. A family of diffeomorphisms $\{f_\varepsilon: 0 \leq \varepsilon < \varepsilon_0\}$, $f_\varepsilon: M_\varepsilon \rightarrow N_\varepsilon$, is called a C^r __family of diffeomorphisms__ if $f^*: M^* \rightarrow N^*$, defined by $f^*(x,\varepsilon) = f_\varepsilon(x)$, is C^r.

For the purpose of reducing the problem to a problem in singular perturbation theory, we introduce the following concept.

__3.3 Definition__ A family $\{\tilde{N}_\varepsilon: 0 \leq \varepsilon < \varepsilon_0\}$ of network perturbations of N is called a __parametrized__ C^r __family of network perturbations__ if the following conditions are satisfied:

(i) The graph $\tilde{G}$ of $\tilde{N}_\varepsilon$ is fixed for every $\varepsilon > 0$.

(ii) If $L_j(i_j,\varepsilon)$ and $C_k(v_k,\varepsilon)$ are the inductance and the capacitance of an extra inductor and capacitor of $\tilde{N}_\varepsilon$, respectively, they satisfy the following. Here we define $L_j(i_j,0) = 0$ and $C_k(v_k,0) = 0$.

(a) $\frac{\partial}{\partial \varepsilon} L_j(x,\varepsilon)|_{\varepsilon=0} \neq 0$, $\frac{\partial}{\partial \varepsilon} C_k(x,\varepsilon)|_{\varepsilon=0} \neq 0$,

for all x,j,k (3.4)

(b) The following functions are C^{r-1}

$$L_j = \begin{cases} \dfrac{\varepsilon}{l_j(i_j,\varepsilon)} , & \varepsilon \neq o \\[2ex] (\frac{\partial}{\partial \varepsilon} L_j(v_j,\varepsilon)|_{\varepsilon=0})^{-1}, & \varepsilon=o \end{cases}$$

$$\tilde{C}_k = \begin{cases} \dfrac{\varepsilon}{C_k(v_k,\varepsilon)}, & \varepsilon \neq o \\[2ex] (\frac{\partial}{\partial \varepsilon} C_k(v_k,\varepsilon)|_{\varepsilon=0})^{-1}, & \varepsilon=o \end{cases}$$

(iii) There is a C^r family $\{F_0(\cdot,\varepsilon): 0 \leq \varepsilon < \varepsilon_0\}$ of functions

$$F_0(\cdot,\varepsilon): \quad C_1(R_0) \rightarrow C^1(R_0)$$

such that the graph of $F_0(\cdot,\varepsilon)$ in $C_1(R_0) \times C^1(R_0)$ coincides with the characteristic manifold $(\Lambda_R)_\varepsilon$ of R_0 in N_ε. Here, for $\varepsilon=0$, $F_0(\cdot,0) = 0$. A similar property is satisfied for $F_\infty(\cdot,\varepsilon)$: $C^1(R_0) \rightarrow C_1(R_\infty)$.

Clearly, $\tilde{N}_0$ is a trivial network perturbation of N.

3.4 Definition: A parametrized C^r family $\{\tilde{N}_\varepsilon: 0 \leq \varepsilon < \varepsilon_0\}$ of network perturbations   of N is   called   a   <u>parametrized</u> C^r <u>family of</u> <u>regularized network perturbations</u> of N, if the configuration manifold $\tilde{\Sigma}_\varepsilon$ of each $\tilde{N}_\varepsilon$ is regular.

A family $\{X_\varepsilon: \; 0 \leq \varepsilon < \varepsilon_0\}$ of vector fields X_ε on C^r manifolds M_ε is called a C^{r-1} <u>family of vector fields</u> if $\{M_\varepsilon: \; 0 \leq \varepsilon < \varepsilon_0\}$ is a C^r family of manifolds and $X_\varepsilon(x)$, $x \in M_\varepsilon$ defines a C^{r-1} vector field on M^*.

3.5 Proposition: <u>Let K be a linear subspace and Λ be a C^r submanifold of</u> $\mathbb{R}^n$ <u>with transverse intersection. Let $\{F_\varepsilon\}$,</u> $0 \leq \varepsilon < \varepsilon_0$ <u>be a C^r family of</u> <u>embeddings</u> $\Lambda \rightarrow R^n$ <u>such that</u> F_0 <u>is the inclusion map and</u> $F_\varepsilon(\Lambda) \pitchfork K$ <u>for</u> <u>every ε. Then, there is a C^r family $\{G_\varepsilon\}$ of embeddings</u> $G_\varepsilon: \Sigma \rightarrow K$, $0 \leq \varepsilon <$ ε_0, <u>such that</u> G_0 = <u>identity and</u> $G_\varepsilon:(\Sigma) = \Sigma_\varepsilon$, <u>where</u> $\Sigma = \Lambda \cap K$ <u>and</u> Σ_ε = $F_\varepsilon(\Lambda) \cap K$.

If G_ε is of class C^{r-1}, then this proposition is well known. When r = 1, this is Proposition 1 of [5]. For the proof of 3.5, the method of [5] is useful. The only fact that we must prove is that if F_ε is C^r close to F_0, then G_ε is C^r close to G_0. For this purpose, the implicit function theorem is useful. If a function $F_\varepsilon(x,y)$ is C^r close to $F_0(x,y)$, then the implicit function $y = f_\varepsilon(x)$ of $F_\varepsilon = 0$ is C^r close to the implicit function $y = f_0(x)$ of $F_0 = 0$.

Remark: When the characteristic manifold Λ_R of N is given such that v_j = $f_j(i_j)$ or i_j = $f_j(v_j)$ for each resistor R_j in R, the existence of a regularized network perturbation $\tilde{N}$ of N is guaranteed by E. Ihrig [3]. In this construction of $\tilde{N}$, $\tilde{N}$ does not contain extra resistors R_0 or R_∞.

Let E be a C^r submanifold of M consisting entirely of equilibrium points of $X_0 \in \{X_\varepsilon\}$. At the point $x \in E$, we have the linear map

$$TX_0(x): \; T_xM \rightarrow T_xM.$$

Since X_0 vanishes identically on E, $TX_0(x)$ induces a linear map

$$QX_0(x): \; T_xM/T_xE \rightarrow T_xM/T_xE$$

on the quotient space.

Let $E_R \subset E$ be an open set for which QX_0 is invertible. E_R is called a _normally regular_ domain of E. For each $x \in E_R$, $T_x E$ has a unique complement N_x which is invariant under $TX_0(x)$. The complement is a realization of $T_x M/T_x E$. Let π_E be the projection:

$$\pi_E: TM|E_R = (TE|E_R) + N \to (TE)|ER.$$

π_E is at least C^{r-2}, since X_0 is C^{r-1}. (In condition (iv) of the Theorem to follow, π_E is C^{r-1}). Let $E_H \subset E_R$ be an open subset for which QX_0 has no pure imaginary eigenvalues. E_H is called a _normally hyperbolic domain_ of E. Let $E_S \subset E_H$ be an open subset where all the eigenvalues of QX_0 have negative real part. E_S is called a _normally stable domain_ of E.

In E_R the _reduced vector field_ X_R is defined by

$$X_R(x) = \pi_E \frac{\partial}{\partial \varepsilon} X_\varepsilon(x)_{\varepsilon=0}. \tag{3.5}$$

Let $f: M \to N$ be a C^r diffeomorphism, $0 \leq r \leq \infty$, and X be a vector field on M. The vector field $f_* X$ on N is defined by

$$(f_* X)(y) = (Tf)X(f^{-1}(y)),$$

where $Tf: TM \to TN$ is the differential of f. Let X,Y be the vector fields on M,N respectively. X and Y are said to be C^s _equivalent_, $0 \leq s \leq \infty$, if there is a C^s diffeomorphism $h:M \to N$ such that h maps each orbit of X onto an orbit of Y. If X is a C^{r-1} vector field on M, then the time rescaled field εX is C^r equivalent to X.

Here we impose the following condition for a regularized network perturbation $\tilde{N}_\varepsilon$ of N.

(E) $\tilde{N}_\varepsilon$ does not contain extra resistors R_0 or R_∞.

3.6 Theorem: _Let_ $\{\tilde{N}_\varepsilon: 0 \leq \varepsilon \leq \varepsilon_0\}$ _be a parametrized_ C^r _family of regularized network perturbation of_ N, $r \geq 2$. _Let_ $\tilde{N}_\varepsilon$ _be the circuit obtained by open-circuiting all resistance branches of_ R_∞ _in_ $\tilde{N}_\varepsilon$ _and short-circuiting all resistance branches of_ R_0 _in_ $\tilde{N}_\varepsilon$. _Let_ Σ, $\tilde{\Sigma}_\varepsilon$ _and_ $\overset{\approx}{\Sigma}$ _be the configuration manifolds_ _of_ N, $\tilde{N}_\varepsilon$ _and_ $\tilde{N}_\varepsilon$, _respectively._ _Then, there is a natural_ C^r _embedding_ $\iota: \Sigma \to \overset{\approx}{\Sigma}$ _and a_ C^{r-1} _family of vector fields_ $\{X_\varepsilon; 0 \leq \varepsilon < \varepsilon_0\}$ _on_ $\overset{\approx}{\Sigma}$ _for sufficiently small_ ε_0, _such that the following properties are satisfied:_

(i) _Let a vector field_ $\tilde{Y}_\varepsilon$ _on_ $\tilde{\Sigma}_\varepsilon$ _be the dynamical system for_ $\tilde{N}_\varepsilon$, $\varepsilon > 0$, _and let_ $\tilde{X}_\varepsilon = \varepsilon \tilde{Y}_\varepsilon$. _Then there is a_ C^r _family of diffeomorphisms_ $\{\psi_\varepsilon: 0 \leq \varepsilon < \varepsilon_0\}$, $\psi_\varepsilon: \overset{\approx}{\Sigma} \to \tilde{\Sigma}_\varepsilon$, _such that_ $(\psi_\varepsilon^{-1}) \circ \tilde{X}_\varepsilon = X_\varepsilon$, $\varepsilon \neq 0$. _If the perturbation_

$\tilde{N}_\varepsilon$ of N <u>does not contain extra resistors</u> R_0 <u>or</u> R_∞ (assumption E), <u>then</u> $\tilde{\Sigma}_\infty$
$= \Sigma$ <u>and</u> $\tilde{X}_\varepsilon = X_\varepsilon$ <u>for each</u> ε.

    (ii) $\iota \Sigma$ <u>coincides with the set of all equilibrium points of</u> X_0.

    (iii)   <u>The normally regular domain of</u> $\iota \Sigma$ <u>coincides with the</u> ι-
<u>image of the regular domain of</u> Σ, <u>i.e.</u> $(\iota \Sigma)_R = \iota(\Sigma_r)$, <u>if</u> $\tilde{N}_\varepsilon$ <u>satisfies</u>
(E).

    (iv)   <u>Let the vector field</u> Y <u>on</u> Σ <u>be the dynamical system of</u> N. <u>If</u>
$\tilde{N}_\varepsilon$ <u>satisfies</u> (E) <u>and</u> (F), <u>then</u> $\iota \circ Y = X_R$ <u>where</u> X_R <u>is the</u>
<u>reduced vector field of</u> $\{X_\varepsilon\}$ <u>on</u> $\iota \Sigma_r = (\iota \Sigma)_r$.

    In the proof, propositions 2.2, 2.4, and 3.5 are used.

4. Application

Our theorem allows dynamical systems describing network perturbations to
utilize the main theorem 9.1 of N. Fenichel's paper [1].

    Let $\{\tilde{N}_\varepsilon: 0 \le \varepsilon < \varepsilon_0\}$ be a parametrized C^r family of regularized
network perturbations of N. In this section we assume (E) in Section 3.

    Hence we denote $\tilde{\tilde{N}}_\varepsilon$ by $\tilde{N}_\varepsilon$. The configuration manifold of $\tilde{N}_\varepsilon$ is fixed,
which is denoted by Σ. The dynamical system of $\tilde{N}_\varepsilon$ for $\varepsilon \ne 0$ is a vector
field Y_ε on Σ, which is determined by:

$$Y_\varepsilon \begin{cases} L_m(i_m)\, \dfrac{di_m}{dt} = v_m, & \text{on } L \\[2ex] C_n(v_n)\, \dfrac{dv_n}{dt} = i_n, & \text{on } C \\[2ex] L_j(i_j,\varepsilon)\, \dfrac{di_j}{dt} = v_j, & \text{on } L^{\cdot} \\[2ex] C_k(v_k,\varepsilon)\, \dfrac{dv_k}{dt} = i_k, & \text{on } C^{\cdot} \end{cases}$$

The dynamical system of N is a vector field Y on Σ, which is determined
by:

$$Y \begin{cases} \dfrac{di_m}{dt} = \dfrac{1}{L_m(i_m)}\, v_m, & \text{on } L \\[2ex] \dfrac{dv_n}{dt} = \dfrac{1}{C_b(v_n)}\, i_n, & \text{on } C \end{cases}$$

The C^{r-1} family $\{X_\epsilon\}$ of vector fields in the Theorem is given by

$$
\begin{cases}
\dfrac{di_m}{dt} = \dfrac{\epsilon}{L_m(i_m)} \; v_m \quad \text{on } L \\[2ex]
\dfrac{dv_m}{dt} = \dfrac{\epsilon}{C_n(v_n)} \; i_n \quad \text{on } C \\[2ex]
\dfrac{di_j}{dt} = \dfrac{\epsilon}{L_j(i_j,\epsilon)} \; v_j \quad \text{on } L^{\backprime} \\[2ex]
\dfrac{dv_k}{dt} = \dfrac{\epsilon}{C_k(v_k,\epsilon)} \; i_k \quad \text{on } C^{\backprime}
\end{cases}
$$

and

$$
\begin{cases}
\dfrac{di_m}{dt} = 0, \quad \dfrac{dv_n}{dt} = 0 \quad \text{on } L \cup C \\[2ex]
\dfrac{di_j}{dt} = \left(\dfrac{\partial}{\partial\epsilon} L_j(i_j,\epsilon)\big|_{\epsilon=o}\right)^{-1} v_j \quad \text{on } L^{\backprime} \\[2ex]
\dfrac{dv_k}{dt} = \left(\dfrac{\partial}{\partial\epsilon} C_k(v_k,\epsilon)\big|_{\epsilon=o}\right)^{-1} i_k \quad \text{on } C^{\backprime}.
\end{cases}
$$

Under a flow on a manifold M, a subset V is said to be <u>locally invariant</u> if there is a neighborhood U of V such that orbit segments which leave V also leave U.

Let K be a compact connected subset of a normally hyperbolic domain Σ_H of $\Sigma_o \subset \Sigma$. Let X be a vector field on $\tilde{\Sigma} \times [0,\epsilon_o]$ defined by

$$
X(x,\epsilon) = (X_\epsilon(x),0) \in T_{(x,\epsilon)}(\tilde{\Sigma} \times [0,\epsilon_o]).
$$

For each $x \in K$, let E_x^s, E_x^u and E_x^c denote the invariant subspaces of $T_{(x,o)}\tilde{\Sigma} \times [0,\epsilon_o]$ associated with the eigenvalues of $TX(x,0)$ in the open left half-plane, in the open right half-plane, and on the imaginary axis, respectively. A manifold W^s is called a <u>stable manifold</u> for X near K if K $\times \{0\} \subset W^s$, W^s is locally invariant under the flow of X, and for all $(x,0)$ $\in K \times \{0\}$, W^s is tangent to $E_x^s \oplus E_x^c$ at $(m,0)$. <u>An unstable manifold</u> W^u and a center manifold W^c are defined in the same way, using $E_x^s \oplus E_x^c$ and E_x^c respectively.

By our Theorem and Theorem 9.1 (i) of [1] <u>there is a</u> C^{r-1} <u>stable manifold</u> W^s, a C^{r-1} <u>unstable manifold</u> W^u, <u>and a</u> C^{r-1} <u>center manifold</u> W^c <u>near</u> K.

The following definitions are also in [1]. Let W^c be a center manifold near K. $W^{ss}(p)$, $p \in W^s$, is said to be a C^{r-1} <u>strong stable manifold</u> near K if

(i) $W^{ss}(p)$ is a C^{r-1} manifold

(ii) $p \in W^{ss}(p)$

(iii) $W^{ss}(p)$ and $W^{ss}(q)$ are disjoint or identical for each p and q in W^s

(iv) $W^{ss}(x,0)$ is tangent to E_x^s at $(x,0)$ for each $x \in K$

(v) $\{W^{ss}(p): p \in W^c\}$ is positively invariant; i.e., $W^{ss}(p).t \subset W^{ss}(p.t)$ fo all $p \in W^s$ and $t \geq 0$ such that $p.[0,t) \subset W^s$. Here .t denotes a flow of X

A <u>strong unstable</u> manifold $W^{uu}(p)$ is defined analogously. By our Theorem and [1] <u>there is a</u> C^{r-2} <u>family</u> $\{W^{ss}(p): p \in W^s\}$ <u>of</u> C^{r-1} <u>strong stable manifolds for</u> W^s <u>near</u> K. If $p \in \sum \times \{\epsilon\}$ <u>then</u> $W^{ss}(p) \subset \sum \times \{\epsilon\}$. <u>Each</u> $W^{ss}(p)$ <u>intersects</u> W^c <u>transversely in exactly one point.</u> <u>Similar properties hold for</u> W^{uu}.

By a theorem of [1], we also have the following. <u>Let</u> $K_s < 0$ <u>be larger than the real parts of the eigenvalues of</u> $QX_o(x)$ <u>in the left half-plane, for all</u> $x \not\in K$. <u>Then there is a constant</u> C_s <u>such that if</u> $p \in W^s$ <u>and</u> $q \in W^{ss}(p)$, <u>then</u>

$$d(p.\tau, q.\tau) \leq C_s \, e^{K_s \tau} \, d(p,q)$$

<u>for all</u> $\tau > 0$ <u>such that</u> $p.[0,\tau] \subset W^s$. <u>Similar properties hold for</u> W^u.

By our Theorem and [1] we have:

$$Y_\epsilon = \frac{1}{\epsilon} X_\epsilon, \quad \epsilon \neq 0$$

$$\iota o Y = X_R = \pi_{\sum_o} (\lim_{\epsilon \to o} Y_\epsilon),$$

and the vector field X_c defined on W^c by

$$X_c(x,\epsilon) = \begin{cases} Y_\epsilon(x) \times \{0\} & \text{if } \epsilon \neq 0 \\ Y(x) \times \{0\} & \text{if } \epsilon = 0 \end{cases}$$

is a C^{r-2} vector field on W^s near K $\times \{0\}$.

By the facts mentioned above, we have the following corollary:

<u>Corollary:</u> <u>Let</u> $\{\tilde{N}_\epsilon: 0 \leq \epsilon < \epsilon_o\}$ <u>be a parametrized</u> C^r <u>family of regularized network perturbations of satisfying</u> (E). <u>Let</u> $\sum$ <u>and</u> $\tilde{\sum}$ <u>be the configuration manifolds of</u> N <u>and</u> $\tilde{N}_\epsilon$, <u>respectively.</u> <u>Let</u> Y <u>and</u> Y_ϵ <u>be the vector fields on</u> $\sum$ <u>and</u> $\tilde{\sum}$ <u>which are the dynamical systems of</u> N <u>and</u> $\tilde{N}$,

respectively. Let $\iota: \sum \to \tilde{\sum}$ be the embedding given by the main Theorem. Let D_S and D_H be compact domains in a normally stable domain $(\iota\tilde{\sum})_S$ and in a normally hyperbolic domain $(\iota\tilde{\sum})_H \backslash (\iota\tilde{\sum})_S$, respectively.

Then, for sufficiently small ϵ, the following holds: If $p \in \tilde{\sum}$ is near D_S, then the motion of p under Y_ϵ jumps to a point near D_S and moves along a trajectory which is arbitrarily close to a trajectory of $\iota * Y$ in $(\iota\tilde{\sum})_S$. If p is near D_H and is not included in the stable manifold W^S near D_H, then p jumps far away along the unstable manifold W^u.

Remark: The assumption (E) is not strong. In fact, to regularize a network by a network perturbation, the inserted resistors R_0 and R_∞ are not effective. This is true since, when the projection $\tilde{\sum} \to \tilde{C}_1 \times \tilde{C}^1$ is a regular map, then $\sum \to \tilde{C}_1 \times \tilde{C}^1$ is regular.

Acknowledgements

I wish to thank T. Matsumoto, M. Vidyasagar, P.R. Bryant and B.C. Haggman. Talking with T. Matsumoto and M. Vidyasagar induced me to study the direction of the present paper. P.R. Bryant and B.C. Haggman supplied me with very helpful suggestions and references during the period in which I prepared this paper at the University of Waterloo.

References

1. N. Fenichel, Geometric singular perturbation theory for ordinary differential equations, J. Differential Equations 31 (1979), 53-98.
2. M. H. Hirsch, C. Pugh and M. Shub, Invariant Manifolds, Lecture Notes in Math., 583 (1977), Springer-Verlag.
3. E. Ihrig, The regularization of nonlinear electrical circuits, Proc. of A.M.S., 47 (1975), 179-183.
4. J. La Salle, Relaxation oscillations, Quart. Appl. Math. 7. (1949), 1-19.
5. T. Matsumoto, G. Ikegami and L. O. Chua, Strong structural stability of resistive nonlinear n-ports, To appear in IEEE Trans. Circuits and Systems, CAS-30.
6. S. S. Sastry and C. A. Desoer, Jump behavior of circuits and systems, IEEE Trans. Circuits and Systems, CAS-28 (1981), 1109-1124.
7. S. Smale, On the mathematical foundation of electrical circuit theory, J. Differential Geometry, 7 (1972), 193-210.
8. F. Takens, Constrained equations; a study of implicit differential equations and their discontinuous solutions, Lecture Notes in Math., 525 (1975), Springer-Verlag, 143-234.

14

On the Dynamics of Iterated Maps III: The Individual Molecules of the M-Set, Self-Similarity Properties, the Empirical n² Rule, and the n² Conjecture

Benoit B. Mandelbrot

IBM Thomas J. Watson Research Center
Yorktown Heights, New York

The definitions of the F^* sets and the M-set of the quadratic map $z \to f(z,\lambda)$ $= \lambda z(1-z)$ are recalled. The M-set subdivides into M-molecules, each of which is shown to be approximately self-similar. To account for this self-similarity, it is conjectured that the derivative of $f(z,\lambda)$ for certain values of z and λ satisfy exactly a certain property called "the n^2 rule". Numerical calculations confirm this rule very accurately, but a mathematical proof is lacking. The fractal dimension of the boundary of each M-molecule is estimated.

0. INTRODUCTION TO A SERIES OF FIVE PAPERS

The present paper is the first of five, which will be called Papers III, IV, V, VI and VII. These papers form a sequel to Mandelbrot 1980, 1982, 1983, which will be called, respectively, Paper I, FGN, and Paper II. Paper VIII, Mandelbrot 1984, continues the present series. The references are collected at the end of Paper VII.

From the mathematical point of view, the main novelty in Paper III resides in a new conjecture stated in Section 6: the "n^2 conjecture". However, the paper also serves as an overall introduction, proposes a convenient vocabulary, describes in new detail several observations that had been sketched in my earlier publications (the "n^2 rule"), and evaluates an interesting fractal dimension.

1. DEFINITION OF THE $F*$-SETS

The successive positions of a dynamic system in discrete time are obtained as the iterates of a suitably defined map. Conversely, any map, such as the quadratic map of the complex plane upon itself, implicitly defines a dynamic system. The study of the dynamics implicit in the iteration of polynomials, of rational functions and of other analytic maps, is dominated by the structure of the attractor and repeller sets; the latter is also called Julia set, $F*$-set (Julia 1918, Fatou 1919). The structure of the $F*$-set is best understood by examining how it can be obtained by successive approximations. This is done in Paper II (Mandelbrot 1983) page 225. Certain special $F*$-sets are discussed further in Papers IV and V of this series.

2. DEFINITION OF THE M-SET

When the map depends upon one or more complex parameters, a further central role is played by the M-set, which was first considered and studied in Paper I, Mandelbrot 1980 (*). According to the simplest definition, (which, however, is not the most widely applicable) the M-set is the set of parameter values for which the $F*$-set is connected.

The present paper focuses on the quadratic map. After suitable change of origin and scale of z, the map $z \rightarrow az^2+bz+c$ depends on a single complex parameter. Depending upon circumstances, the map is best written either as $z \rightarrow f(z,\lambda)=\lambda z(1-z)$ or as $z \rightarrow f*(z,\mu) = z^2-\mu$, where $\mu = \lambda^2/4-\lambda/2$. The point at infinity is a stable fixed point of this map for all values of λ(resp., of μ), and the Julia set $F*_\lambda$ (resp., $F*_\mu$) is best visualized as the boundary of the set of z-points that fail to iterate to infinity under the action of the map f(resp., f*).

As to the M-set M_λ (resp., M_μ), it is best constructed as the set of values of λ (resp., μ) such that the critical value of z, which is z=1/2 (resp. z = 0), fails to iterate to infinity under f (resp., f*).

To implement this definition, the first step is to determine a suitable neighborhood of infinity. Paper II p. 226 shows that, in the case of M_μ, the disc of radius 2 is such a neighborhood. That is, the condition $|f_k*(0,\mu)|>2$ on the k-th iterate of f* is a sufficient condition

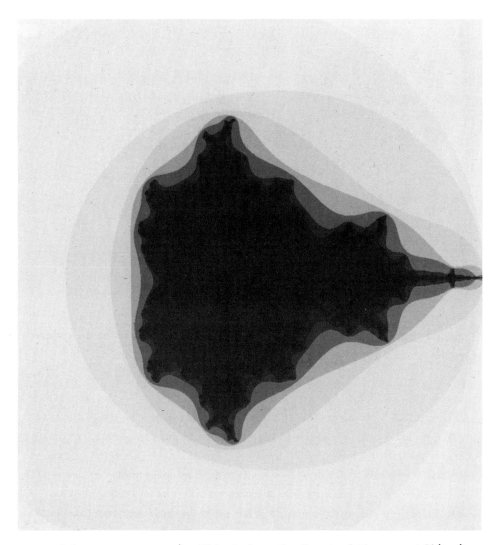

Figure 1 (parameter space). This is how the M-set of the map $z \to f*(z,\mu) = z^2-\mu$ is constructed as the limit of monotone decreasing approximations. Each approximation is bounded by a lemniscate $|f_k*(0,\mu)| = 2$, several of which are shown here as the boundaries between different shades of gray. The values of k are 1, 2, 3, 4, 5, 7, 9 and 14. The last 3 domains being hard to distinguish here, a more distinct view is shown in Figure 2.

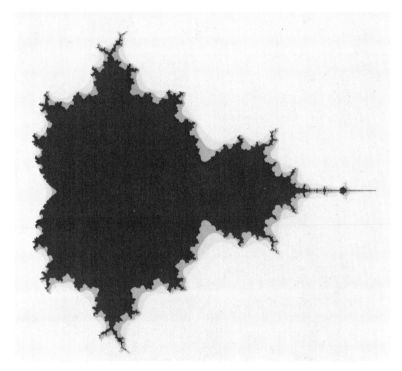

Figure 2 (parameter space). The last three approximations in Figure 1, for k = 7,9 and 14, are seen more distictly than in Figure 1.

for z = 0 to iterate to infinity under f*. Thus, successive approximations circumscribed to M_μ are yielded by the sets M_μ^k defined by $|f_k*(0,\mu|<2$. These sets are illustrated on Figures 1 and 2, which play the same role for M_μ as the Figures on page 225 of Paper II play for $F*$.

The equation $|f_k*(0,\mu)| = 2$ defines an algebraic curve called "lemniscate". The point $\mu = 2$ belongs to all these leminiscates. All other points of the lemniscate of order k'>k are entirely contained within the lemniscate of order k. For k = 1, this lemniscate is precisely a circle. Increase of k results in increasingly fine detail appearing in M_μ^k.

Nevertheless, numerical observation showed that all these lemniscates continue to be topologically identical to circles: they are Jordan curves or simply connected curves. This is a significant observation, because for all z ≠ 0 the lemniscate $|f_k*(z,\mu)| = 2$ eventually breaks down into a collection of an increasing number of topological circles. Since the sets $|f_k*(0,\mu)|\leq2$ are increasingly close circumscribed aproximations to the M_μ

set, the numerical evidence that the approximations are connected suggested that the M_μ set is also connected. An actual proof of this conjecture was beyond my powers, but was achieved by Douady and Hubbard 1982 (see also Douady 1983).

High-k approximations of the sets M_λ and M_μ are shown in FGN (Mandelbrot 1982) pp. 188 and 189, and an attempt is made to reproduce them as Figures 3 and 4 of this paper.

3. THE DIVERSE COMPONENTS OF THE M-SET.

The structure of the M-set is extremely rich. To make it easy to remember, I keep refining the terminology, based on geography, chemistry and physics, used in Papers I and II.

A) The M-molecules. A large "continental molecule" is clearly visible on Figures 3 and 4, and it is surrounded by a denumerable infinity of "island molecules" whose existence I discovered. In printing, the islands tend to come out as indistinct specks (or even to be cleaned off by overzealous printers); however they are very important to the theory of iteration.

Figure 3 (parameter space). A tight approximation of the M-set of the map $z \rightarrow z^2-\mu$, using the lemniscate $|f_k*(0,\mu)| = 2$, with a very large value of **k**.

Figure 4 (parameter space). A tight approximation of the M-set of the map $z \to \lambda z(1-z)$, using the lemniscate $|f_k(1/2, \mu)| = $ constant, with a very large value of k. Figures 3 and 4 deduce from each other by the transformation $\mu = \lambda^2/4 - \lambda/2$. For the purpose of understanding the "n^2 rule" introduced in this paper, the parameter λ is the more perspicuous one.

Each island molecule's shape is like that of the continental molecule of M_μ, up to slight deformation. See Paper II, p. 227.

B) The M-atoms. Each M-molecule decomposes into a denumerable infinity of "M-atoms". An M-atom is a maximal open domain of M's such that the corresponding F^*-sets are topologically identical. Atom boundaries are algebraic curves, with a well-defined tangent at every point, except that some atoms are cardioids with a single cusp.

C) The bonds. The boundaries of two M-atoms within the same M-molecule may share a point called "bond". The boundaries of two bonded atoms share a common tangent at the bond.

D) Bifurcation and Confluence. A bond can be crossed in either of two opposite directions, to be referred to as "upstream" and "downstream". By definition, a crossing in the upstream direction provokes a bifurcation of the attractor of the map f(z), and a crossing in the downstream direction leads to the contrary of bifurcation, which I labeled "confluence". One can move within a molecule from any other atom via a finite sequence of confluences and/or bifurcations.

A sequence of bifurcations can continue ad infinitum, but a sequence of confluences stops necessarily at an atom called the molecule's "large atom". The continental molecule of M_μ (Figure 1) includes two equal large atoms, which are exact discs. For the continental molecule of M_μ (Figure 2) and for all island molecules of M_μ and M_λ, the large atom is a cardioid. All other atoms are discs or near-discs.

E) The M-nuclei. Each atom contains a "nucleus", defined as being a point such that the attractor of the corresponding $f(z,\lambda)$(resp.,$f(z,\mu)$) contains the critical point $z = 1/2$ (resp., $z = 0$). James A. Given has written a computer program to seek the nuclei numerically.

F) The M-radicals. Any disc-shaped atom together with all the atoms upstream of it will be said to form an "M-radical". Thus, the continental molecule of M_λ is made of two radicals, but the continental molecule of M_μ is not a radical. A radical has a single free bond, hence is "univalent". Each radical contains subradicals, subsubradicals, etc. ad infinitum.

G) The semi-open M-set. It is defined as the union of the (open) atoms and of the inter-atomic bonds. The semi-open variant of the M-set (first used in Paper I) is needed in subsection H, and it continues to be defined for maps to which the above-given definition of the M-set fails to extend.

H) The closed M-set and its "devil's polymer" structure. The fact
that the M-set is connected implies that all the molecules are also linked
to each other, but the links between the M-molecules are very different
from the links between the M-atoms. Take the nuclei of two molecules, A
and B, and link them within M_μ or M_λ by a continuous curve parameterized
by $u \epsilon [0,1]$, in such a way that the u's contained within a molecule form
one open interval. Two such intervals never touch, and the complement of
these intervals' union is a fractal dust. By analogy with the widely used
term "devil's staircase", the molecules crossed between A and B will be
said to form a "devil's causeway", and the M-set will be said to have a
"devil's polymer" structure. In the present case of the quadratic map,
this is a branched polymer that has no closed loops.

I) The "polymer glue". The difference between the full closed M-set
and the semi-open M-set is made of the scattered points that "glue" the
molecules together. These points form a very small set: a fractal dust.

J) Chaos. When λ or μ belong to the "polymer glue", the attractor
of $f(z,\lambda)$ or $f(z,\mu)$ is not a finite cycle, but is either a fractal dust or
a collection of intervals. It has become customary to refer to the
corresponding maps as "chaotic". Much attention has been given to the
finding that a very simple map can be chaotic.

When the map depends on a parameter, it is interesting to know
whether or not chaos is a "prevalent" phenomenon, i.e., whether or not a
randomly chosen map has a positive probability of being chaotic.

When z and μ are restricted to the real axis (the facts are
summarized in Collet and Eckman 1980), the "glue" has positive linear
measure, hence chaos is a prevalent phenomenon.

In the case of complex z and μ, to the contrary, there is every
indication that the glue has a vanishing areal measure. This should mean
that "chaos" is <u>not</u> a prevalent phenomenon. Furthermore, whenever μ_0 lies
in the polymer glue, the z_0 that behave in orderly fashion iterate to ∞.
The set of points that fail to iterate to ∞ is smaller than it is for any
of the nonchaotic μ's that can be found arbitrarily close to μ_0. In other
words, we have the interesting observation that "chaos on the real
interval" implies "maximal order in the complex plane".

4. OUTLINES OF PAPERS III to VII

Papers III to IV concern the structure of the M-set, and Papers VI and VII
concern the structure of some F-sets of special interest.

The remainder of **Paper III** concerns the shape of the radicals of the M-set. It reports certain approximate scaling relations I have observed in them, and argues that the reasons behind these relations lies in a conjectured "n^2 rule".

These relations´ validity extends beyond the quadratic map, as I have found that very similar shapes are the building blocks of the M-sets of a broad class of other maps, for example of the map $z \rightarrow \lambda(z-1/z)$(see the second and later printings of FGN, pp. x and 465, and Paper VIII). Paper I ends with comments on the fractal dimension D of the radical or molecule boundaries. This dimension D is the same for the specks´ and the continent´s boundary, in fact is a quantity having a wide class of universality.

Paper IV concerns a shape R called "reduced radical". This shape is defined as satisfying exactly the scaling relations that Paper III claims are approximately valid for the radicals of the M-set. Curious number-theoretical considerations enter in.

Paper V concerns the "devil´s polymer" structure of the whole M-set. It is conjectured that this structure is so dense that the boundary of M has a fractal dimension equal to 2.

Paper VI concerns certain Julia sets whose structure is a "devil´s polymer" reminiscent of the M-set, and conjectures that the corresponding atoms´ boundaries are smooth curves, of fractal dimension 1.

Paper VII concerns the Siegel discs, which are interpreted as Julia sets obtained as boundaries of sequences of plane-filling ("Peano") Julia sets.

Paper VIII (which is not in this book, see Mandelbrot 1984) tackles some issues of chaos in the case of the map $z \rightarrow \lambda(z+1/z)$.

5. EMPIRICAL OBSERVATIONS ON THE RADICALS OF M_λ

Consider the left-hand-side half of the set M_λ of Figure 1. It is a circle surrounded by sprout-shaped aggregates of atoms, "M-radicals" in my chemical terminology, whose roots are the points of the form $\lambda_0 = \exp(2\pi i m/n)$. The (open) atom rooted at the bond λ_0 is the locus of the points that satisfy the condition $|f_n´(z_\lambda,\lambda)| < 1$, where z_λ is one of the values in the stable periodic cycle of n points corresponding to λ. The distance between λ_0 and the atom´s nucleus will be called the atom´s "radius", and denoted by $r_{m,n}$.

First Observation. The radicals are near identical in shape to one half of M_λ itself. Hence the same is true of the sub-radicals of any order, and one half of M_λ is nearly self-similar.

Second Observation. Taking m/n to be an irreducible fraction, the radius $r_{m,n}$ is nearly independent of m.

Third Observation in Rough Approximation (Mandelbrot 1983). The radius $r_{m,n}$ is nearly equal to n^{-2}.

Corollary. Compare the radicals whose roots result from different sequences of bifurcation, but with orders having the same product $n_1,n_2...n_k$. Combining the above observations predicts that these sprouts have nearly identical radii.

Third Observation in a More Refined Approximation. The quantity n^{-2} gives a close approximation to the distance from the bond λ_0 to the projection of the nucleus upon the half-line from 0 to λ_0.

6. A MATHEMATICAL CONJECTURE WHOSE VALIDITY WOULD ACCOUNT FOR THE OBSERVATIONS IN SECTION 2.

Behind the terms "nearly identical", "nearly independent" and "nearly equal" in the preceding section, lurks significant variability, to be explored in a forthcoming work. Nevertheless, these observations suggested the following precise mathematical statement. It has been verified numerically for a large number of values of m/n, but I have no general proof for it.

The "Special n^2 Conjecture". Consider the half-line that radiates from $\lambda = 0$ to the bond $\lambda_0 = \exp(2\pi im/n)$, which corresponds to bifurcation from 1 to n. Take the derivative of $f_n^{'}(z_\lambda,\lambda)$ along this half line. At the point λ_0, this derivative is equal to $-n^2$.

Generalization. The function g(λ). The special n^2 conjecture concerns derivatives along a line that is orthogonal to the boundaries of the atoms that are bonded at λ_0. Furthermore, if $|\lambda| < 1$, the function $|f'(z_\lambda,\lambda)|$ is identical to $|\lambda|$; hence, its derivative along the same radial direction is 1. Now define the function g(λ) for λ is the semi-open M-set, as equal to $f_n^{'}(z_\lambda,\lambda)$ where n is the period of the limit cycle corresponding to λ and z_λ is a point in that limit cycle. Clearly, $|g(\lambda)|$ = 1 when λ is a bond. Restated in terms of g(λ), the special n^2 conjecture asserts that when λ crosses certain special points of bifuration into n, the gradient of $|g(\lambda)|$ is multiplied by $-n^2$. This restatement is clearly invariant, in the sense that it remains unchanged if the parameter λ is replaced by μ or by any of a wide range of functions of λ that are sufficiently smooth near λ_0 and preserve orthogonality. Furthermore, the invariant form lends itself directly to the following generalization.

The "General n^2 Conjecture". Let λ_0 be a bond between two atoms, corresponding to bifurcation from order n_0 to order $n_0 n$, and consider $g(\lambda)$ along a curve that crosses λ_0 in a direction orthogonal to the bonded atoms' boundaries. The ratio of the derivatives of $|g(\lambda)|$ to the right and the left of the point λ_0 is $-n^2$.

Comment. The observations in Section 5 would hold true if $f_n'(z_\lambda,\lambda)$ were a linear function of λ within each atom, and the n^2 conjecture were confirmed. It would follow indeed that every atom in the continent of the M_λ-set would be a disc, and that the function $f_n'(z_\lambda,\lambda)$ would vanish at the nucleus and be equal to 1 in modulus along the disc's circular boundary. The atom's radius would be the inverse of the derivative of $f_n'(z_\lambda,\lambda)$ at the root of this atom, hence would be independent of m and equal to n^{-2}.

7. <u>ESTIMATE OF THE FRACTAL DIMENSION OF THE M-MOLECULE'S BOUNDARY.</u>

The boundaries of the M-atoms are algebraic curves, called "lemniscates", hence are very smooth, of fractal dimension D=1. However, the boundaries of the M-molecules are each composed of an infinity of smooth boundaries of atoms. When a fractal is the union of smooth curves (FGN, p. 118), fractal dimension is not a measure of the roughness of a shape but of its fragmentation. When we study the normalized radical R in Paper IV, we shall find that its dimension is the positive solution of the D-generating equation $\Sigma \phi(n)n^{-2D} = 1$, where the sum is from 2 to ∞ and $\phi(n)$ is Euler's number-theoretical function. Numerically, D = 1.239375. This value also serves as an estimate of the dimension of the boundary of a M-molecule. How precise is this estimate is not known. A few hybrid D-generating equations gave estimates that are close to the above one, but of uncertain reliability.

ACKNOWLEDGMENTS. I acknowledge numerous useful discussions with V. Alan Norton and James A. Given. The computer programs used to draw Figure

1 and 2 were written by Mark R. Laff and V.A.N.   The programs used to evaluate the radii $r_{m,n}$ were written by J.A.G., under the guidance of V.A.N.   The assistance of Linda Soloff and of Janice H. Cook was valuable.

ACKNOWLEDGMENTS.   I acknowledge numerous useful discussions with V. Alan Norton and James A. Given.   The computer programs used to draw Figure 1 and 2 were written by Mark R. Laff and V.A.N.   The programs used to evaluate the radii $r_{m,n}$ were written by J.A.G., under the guidance of V.A.N.   The assistance of Linda Soloff and of Janice H. Cook was valuable.

(*)**Editors' Note:**   The accepted term for the M-set is "Mandelbrot set".

Note Added in the Second Printing

A)   Figures 1 and 2 lend themselves to spectacular color renderings.   The best at present are in Peitgen and Richter 1985 (See page 253).

B)   The Conjecture on p.   223 has now been proven to be correct, by John Guckenheimer of Cornell University and Richard McGehee of the University of Minnesota.   See A Proof of Mandelbrot's N^2 Conjecture.   Report of the Mittag-Leffler Institute. Djursholm, Sweden, 1984.

15

On the Dynamics of Iterated Maps IV: The Notion of "Normalized Radical" R of the M-Set, and the Fractal Dimension of the Boundary of R

Benoit B. Mandelbrot

IBM Thomas J. Watson Research Center
Yorktown Heights, New York

A "normalized radical" R of the M-set is defined as the shape that satisfies exactly all the approximate self-similarity properties of the M-set molecules the quadratic map, as reported in Paper III. Explicit constructions show that the complement of R is a "σ-lune", and prove that does not self-overlap. The fractal dimension D of the boundary of R is shown to satisfy $\sum_{2}^{\infty}\phi(n)n^{-2D} = 1$, where $\phi(n)$ is Euler's number-theoretic function. The solution is D = 1.239375. In a rough first approximation, D is the solution of the equation $\sum_{2}^{\infty}n^{1-2D} = \zeta(2D-1) - 1 = \pi^2/6$, where ζ is the Riemann zeta function. The same D applies to the M-sets of other maps in the same class of universality. Some interesting "rank-size" probability distributions are introduced.

1. INTRODUCTION

Paper I, published in 1980, was the first to investigate the structure of a remarkable set in the complex plane, called the M-set(*), which is defined in Paper III above, and plays a central role in the dynamics implicit in the iteration of polynomials, of rational functions and of

other analytic maps. Like paper III, this paper focuses on the quadratic
map, written as z→f(z,λ) = λz(1-z), where z and λ are complex numbers.
The molecules of the corresponding M-set were shown in Paper III to be
approximately self-similar. Section 2 defines a "normalized radical" R,
as being a shape for which these self-similarity rules hold exactly, and
asserts that R fails to self-overlap. Section 3 proves that R fails to
self-overlap, via an explicit construction of the complement of R.
Section 4 is an aside into number theory, needed in Section 5. Section 5
evaluates the fractal dimension of the boundary of R. Section 6
introduces some interesting probability distributions.

2. THE NORMALIZED M-RADICAL R. DEFINITION, ABSENCE OF SELF-OVERLAP

Definition of R as a σ-disc. The normalized M-radical R, Figure 1, is a
σ-disc, that is, a denumerable collection of closed disc-shaped atoms, and
is constructed recursively as follows. The 0-th generation atom is the
unit disc, the 1-st generation atoms are discs tangent to the unit disc,
the points of tangency ("bonds") being of the form λ = exp(2πim/n) and the
radii being equal to n^{-2}. Each k-th generation atom of radius n^{-2} is
given intrinsic coordinates such that its center is of coordinate 0 and
its bond to the (k-1)th generation disc is of coordinate-n^{-2}. Then this
k-th generation atom is "decorated" by (k+1)st generation atoms whose
bonds´ intrinsic coordinates are of the form $n^{-2}\exp(2\pi im/\nu)$ and whose
radii are equal to $n^{-2}\nu^{-2}$.

Unbounded variant of the radical. A variant of this construction
alternates stages of interpolation and extrapolation. The process is
tedious to write down, but the result is in effect the same as the limit
of simple interpolation (as above) that has been blown up in the ratio 4^{∞},
while its right-most point remains fixed. A piece of this extrapolation
is seen on Figure 1, using a partly-filled disc.

Definition of the term, "lune". Given a circle $C´$ and a circle C''
that is tangent to $C´$ and otherwise contained in $C´$, the set of points
inside $C´$ and outside C'' is called an open lune.

Representation of the complement of R as a σ-lune. The complement of
R is covered by a σ-lune, namely by a denumerable infinite collection of
open lunes. This construction results from various properties of
R embodied in the lemmas of Section 3. An advanced stage of construction
is illustrated on Figure 2. The construction is so effective that Figure

Figure 1. By definition, the normalized radical R of the M-set is a σ-disc. Here it is approximated by the union of a fairly large number of closed discs.

2 is not very telling. Therefore, an alternative illustration is given in Figure 3, wherein each lune is represented by a fan of lines, and it is clearly seen how the various fans overlap. (It is amusing to report that this attractive and illustrative Figure 3 had originally resulted from an error of computer programming.)

    Corollary. Property of non-self-overlap. The closed discs constructed as in the above definition of R do not overlap except at their bonds.

    A map, $z \to h(z, \eta)$, with η a complex parameter, such that its M-set is made of normalized M-radicals R; the dependence of η on μ is singular.

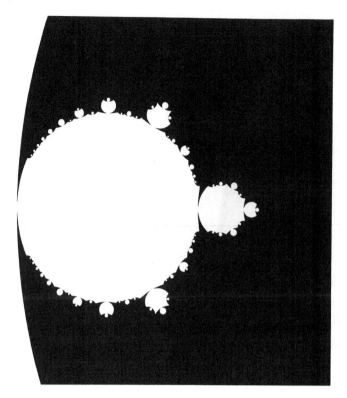

Figure 2. The complement of the normalized radical R is shown in this paper to be a σ-lune. Here it is approximated by the union of a not-too-large number of lunes. By definition, the subradicals are reduced-scale versions of the radical itself. Therefore, the examination of subradicals of increasing size shows how an increasing number of lunes defines the complement of R: first as a disc, then in increasing detail.

Denote by η the complex variable in the plane in which R is embedded. The interiors of R and the continental M-molecule of $f*(z,\mu) = z^2-\mu$ can be put in a doubly continuous one-to-one correspondence $\mu(\eta)$ in such a way that for $h(z,\eta) = f*[z,\mu(\eta)]$, the $g(\lambda)$ function (Paper III, Section 6) is piecewise linear within each M-atom. Thus the M-molecule of $h(z,\eta)$ is R itself. Recall, however, that Feigenbaum "universality", as applied to the map f* restricted to the real axis, tells us that the ratio of the diameters of successive atoms intersected by the real axis takes on the value 4.66920 for a wide class of maps. Since for the map h(z,η) its

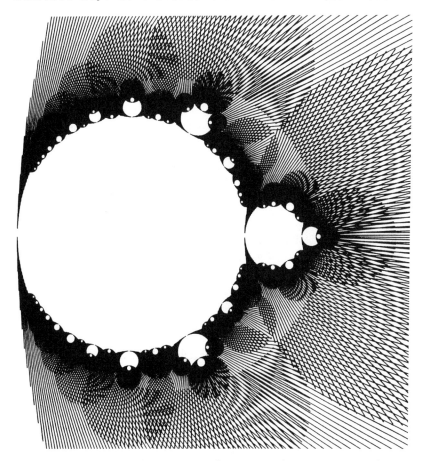

Figure 3. Identical to Figure 2, except for a programming error: Each
lune should have been drawn as the union of filled-in quadrangles, but is
drawn here as the union of these quadrangles' outlines. This Figure
explains the σ-lune algorithm better, and it is attractive.

value is 4, $h(z,\eta)$ does not belong to the same class of universality as
$f*(z,\mu)$.

Indeed, consider as example the neighborhood of the limit points μ_∞
and η_∞ corresponding to an infinite sequence of 2-bifurcations. In this
neighborhood, $(\mu_\infty-\mu) \propto (\eta_\infty-\eta)^{\log 4.66/\log 4}$. Similarly, the counterparts of
λ for series of n-bifurcations, with n>2, were evaluated by Cvitanovic and
Myrheim 1982; they too are close to, but different from, n^{-2}. Thus, the
correspondence $\mu(\eta)$ is very singular on the boundary of R.

Deflated form of R. Figure 4 shows what would happen to R after each
disc of perimeter $2\pi n^{-2}$ is flattened into a "doublet", namely an interval

Figure 4. The boundary of R, after each of its discs has been "deflated" into two superposed segments. The segments corresponding to 2-bifurcations are aligned, and those corresponding to higher bifurcations are orthogonal.

of length πn^{-2} covered in one direction then in the opposite one. The right side is parameterized by θ's in $[0,\pi]$, and the left side by θ's in $[\pi,2\pi]$. Two atoms tangent at $\theta = \pi$ are replaced by doublets colinear to the originators and all other orthogonal atoms are replaced by doublets orthogonal to the original one.

3.   CIRCLES THAT OSCULATE R, AND PROOF THAT THE COMPLEMENT OF R CAN BE COVERED BY A σ-LUNE.

Comment. The lemmas in this section have obvious approximate counterparts in the actual M-set of the map $z \rightarrow \lambda z(1-z)$.

Lemma A. An infinite sequence of bifurcations leads to the point $[-5/3,0]$. Proof. The diameters of the discs in this infinite sequence are 2, $2(1/4)$, $2(1/4)^2$; etc. These diameters' sum is 8/3.

Lemma B R is contained in the closed disc of center 0 and radius 5/3. Proof. By Lemma A, this disc contains all the bonds that obtain by a finite sequence of bifurcations or order 2. It suffices to consider a sequence of bifurcations, such that the orders $n_1, n_2 \ldots n_g$, are not all equal to 2, and to show that it always leads to a bond that lies within the above disc. Indeed, consider the broken line that goes from 0 to the first bifurcation bond, then to the center of the corresponding atom, then to the next bifurcation bond, etc. ... Each vector in this broken line is at most equal in modulus to that of the line which corresponds to $n_g \equiv 2$, and at least one of these vectors is smaller in modulus. Therefore, those vectors' sum is less than 5/3.

Lemma C. Near its root at $\exp(2\pi i \theta)$ with $\theta = 0$, R is locally osculated by a circle of curvature (inverse radius) equal to $\rho^{-1} = 1 - 4/3\pi^2$. Proof. Apply Lemma B to each subradical of R. A subradical whose bond coordinates are $\exp(i\theta) = \exp(2\pi im/n)$ lies within a disc whose radius is $(5/3)n^{-2}$ and whose center is at a distance of $1 + n^{-2}$ from 0. A fortiori, it lies in a disc whose radius is $(5/3)(m/n)^2$ and whose center is at a distance $1+(m/n)^2$ from 0. These enlarged bounding discs badly overlap, but their union's outside boundary is a smooth curve. Near $\theta =$ 0, this boundary is the curve of equation $x = -y^2/2 + (8/3)(y/2\pi)^2 = -y^2[1/2-2/3\pi^2] = y^2/2\rho$. Hence its local curvature is $\rho^{-1} = 1-4/3\pi^2$, meaning that the osculating circle has the radius $\rho{\sim}1.1562$.

Lemma D R is contained in disc of center $- 1/3$ and radius $4/3$. Proof. Writing $\theta/2\pi = u$, the distance from $- 1/3$ to a point of an enlarged bounding disc is bounded by the quantity $(1/3)\{5u^2+\sqrt{[1+9(1+u^2)^2 + 6(1+u^2)\cos(2\pi u)]}\}$. This quantity is readily seen to be $4/3$ for $u = -1,0$ or 1, and to be smaller than $4/3$ for all other u in $[-1,1]$.

Lemma E.   Near the bond point $\exp(2\pi im/n)$, divide R into a subradical, whose large atom is a circle of radius n^{-2}, and a remainder. The subradical is locally osculated by a circle of curvature $n^2(1-4/3\pi^2)$, and the remainder is locally osculated by a circle of curvature $4n^2/3\pi^2-1$.

Corollary of Lemma E.   Osculating lunes. The subradical and the remainder do not overlap locally; in fact, they are locally separated by the "osculating lune" contained between the osculating circles.

Comment on Lemma E. Globally, each osculating lune intersects R. Therefore, covering of the complement of R requires lunes that are smaller than the osculating lunes.

Lemma F.   Complementary lunes. To each subradical of R whose big atom is a disc of radius r, attach the "complementary lune" that is contained between the following two circles: the atom's circumscribed circle, whose radius is $(4/3)r$, and the circle of radius $(6.6)(4/3)r$ that is tangent to the circumscribed circle at the bond of the subradical. This lune is called complementary because it lies outside R. The intersection of the exterior of R with the whole radical's circumscribed circle is covered by "σ-lune" defined as the union of the subradicals' complementary lunes.

Comment on Lemma F. One of the circles that bound this lune is obvious, since in order for a lune to lie outside of R, it is necessary and sufficient that one of the bounding circles be circumscribed to the atom. As to the numerical factor 6.6, it was obtained by trial and error. However, it can easily be proven that the exterior of R can be covered by slightly more involved sets.

As seen on Figure 3, the structure of the exterior of R is well outlined by a small number of lunes. The same structure is also clearly visible on one half of the M-set L. It has motivated the present investigations, and is explained by it.

4. EULER'S NUMBER-THEORETICAL FUNCTION $\phi(n)$ AND KIN.

The number of irreducible fractions of denominator n is Euler's number-theoretical function $\phi(n)$. Define $\phi*(n) = \sum_{u=1}^{n} \phi(u)$. By an old theorem of Mertens (Hardy and Wright 1960, Theorem 330), $\phi*(n) \sim (3/\pi^2)n^2 + O(n\log n)$ as $n \to \infty$, hence the rough estimate $\phi(n) \sim (6/\pi^2)n \sim .6079271\ n$. A more detailed investigation that I carried out suggested the more precise representation $\phi*(n) = (3/\pi^2)\{n^2+n[1+\beta(n)]+\alpha\}$. This formula defines a number α, and a function $\beta(n)$ that turns out to look very much like a stationary random function of n that oscillates around zero. The "random term" $n[1+\beta(n)]$ in this representation is of the order of magnitude of the square root of the "drift term" n^2+n. The function $\beta(n)$ has very interesting properties that will be reported elsewhere in greater detail. Observe that $(6/\pi^2)n$ is the finite difference of the above $\phi*(n)$ with $\beta(n)$ set to 0.

5. VALUE OF THE FRACTAL DIMENSION D OF THE BOUNDARY OF R.

By the self-similarity property of R, D is the solution of the D-generating equation $\sum_{n=2}^{\infty} \phi(n)n^{-2D} = 1$. A numerical solution obtained by Newton's method is D = 1.239375. Using the leading term of the Mertens formula for $\phi(n)$, one obtains a simple approximate D-generating function, namely $(6/\pi^2)\sum_{n=2}^{\infty} n^{1-2D} = 1$, that is $\zeta(2D-1) = 1+\pi^2/6$. The numerical solution is D = 1.245947. This small D explains why Figure 4 is so "skinny".

6. RANK-SIZE PROBABILITY DISTRIBUTIONS SUGGESTED BY THE SUBRADICALS OF R.

Rank the subradicals directly attached to R by increasing values of n and, for each n, by increasing values of m. The rank of a subradical in this list will be denoted by ρ. The number of subradicals whose base circle is of radius n^{-2} is Euler's $\phi(n)$ (Section 3); hence $\sum_{1}^{n} \phi(u) = \phi^*(n)$ is the number of ρ's such that base circle radius is $\geq n^{-2}$. In the Mertens approximation, $\phi^*(n) \sim (3/\pi^2)n^2$; hence we find the rules

$$a \text{ subradical's linear size} \sim 1/\text{rank}$$
$$a \text{ subradical's area} \sim 1/(\text{rank})^2$$

These are examples of the so called "statistical rank-size" rule, which is followed by many natural phenomena, of which a few are mentioned in FGN (e.g., Chapter 38). Unfortunately, examples where this rule is obtained by a theoretical argument are comparatively rare, which adds value to the present example. Linear size cannot be weighted to yield a probability, because $\Sigma(1/n) = \infty$. But area can be weighted in this way: it suffices to divide it by the sum of all subradicals' areas.

Does there exist an exponent D such that n^{-D} is the probability of a subradical, without need of a weighting prefactor? Such a D must satisfy $\sum_{n=2}^{\infty} \phi(n)n^{-2D} = 1$, hence it is the fractal dimension of the boundary of R. Analogy with other nonrandom fractals suggests that the Hausdorff measure of the boundary of R is positive and finite and can be taken to be 1. If so, n^{-2D} (without any prefactor) is the Hausdorff measure of the boundary of the subradical rooted at $\exp(2\pi im/n)$. This measure gives mathematical meaning to the intuitive notion of points distributed uniformly on the boundary of R.

7. THE INTERESTING NUMBER-THEORETICAL FUNCTION $\nu(n)$.

Denote by $\nu(n)$ the number of circles of radius $\geq n^{-2}$ contained in R. For many fractal curves whose complement is contructed as a union of open "gaps" (for example, for the Apollonian gasket; FGN, p. 170) the dimension is known to be the value of D that rules the distribution of the linear sizes of the gaps. That is, F being a numerical prefactor, $\nu(n) \sim F(n^{-2})^{-D} = Fn^{2D}$. It is reasonable to assume (but remains to be proven) that this relation also holds for R. This conjecture suggests that $\nu(n) \propto n^{2D}$.

The function $\nu(k)$ has a direct arithmetical interpretation. It is the number of distinct products of irreducible fractions, such that the product's denominator is n, being granted that each permutation of

distinct multiplicands is counted separately. This definition would look
contrived, were it not for the application that motivated it. One
wonders, inevitably, whether or not this $\nu(n)$ plays any other role in
"non-abelian" arithmetic.

ACKNOWLEDGMENTS. I acknowledge numerous useful discussions with V. Alan
Norton, James A. Given and Janice H. Cook. The computer programs used to
draw Figures 1, 2 and 3, and to estimate D, were written by H.J.C.

References: See the end of Paper VII

16

On the Dynamics of Iterated Maps V: Conjecture That the Boundary of the M-Set Has a Fractal Dimension Equal to 2

Benoit B. Mandelbrot

IBM Thomas J. Watson Research Center
Yorktown Heights, New York

It is conjectured that the boundary of the M-set of a complex map (Paper I(1980) and Paper III of this series of papers) has a fractal dimension equal to 2.

The ambition of the present paper is to help understand the "devil's polymer" structure that binds together the molecules of the M-set. The M-set may be a branching polymer (tree), as for the map $z \rightarrow \lambda z(1-z)$, in which case the complement is connected. It may be a net whose complement is a collection of open sets. And it may be a net whose complement's structure remains as yet undetermined, as for the map $z \rightarrow \lambda(z+1/z)$. The present text is devoted to $z \rightarrow \lambda z(1-z)$, but the conjecture it describes is surely of broad applicability.

Conjecture. The boundary of the M-set is a curve of fractal dimension equal to $D = 2$.

Background. Fig. 1, a negative of Plate 164 of Mandelbrot 1982, results from a recursive creation of (white) branches. The ratio of branch lengths after and before each branch point increases slowly to $\sqrt{2}/2$

235

Figure 1. Construction of two curves of fractal dimension D = 2; from FGN
p. 164.

as one moves towards the branch tips, and the ratio of width to length
decreases to zero. On the left side of the illustration, this
width/length ratio decreases even faster than on the right side. If one
zooms in, the picture remains unchanged in overall shape, but the relative
thicknesses of similarly positioned branches decrease to zero. This means
that these diagrams were so designed that they fail to be self-similar.
This is how both achieve the desired result that the fractal dimension of
the branch sides and the branch tips is D = 2. (The rate at which the
ratio width/length decreases to zero does not affect the value of D.)

Evidence. Figures 2 and 3 describe the local structure of the
boundary of the M-set by "zooming" onto the neighborhood of a typical
point of the boundary of the continental molecule. More precisely, σ
being the "golden ratio" $\sigma = (\sqrt{5}-1)/2$, Figures 2 and 3 inspect in
increasing detail a small piece of the M-set near $\lambda = \exp(2\pi i \sigma)$, and a
smaller sub-piece. Each piece is roughly centered on a parameter value of
the form $\lambda_m = \exp(2\pi i \sigma_m)$, where σ_m is the ratio of the m-th and the
(m+1)st Fibonacci numbers. These σ_m are successive approximations to σ.
And the size of each piece is such as to contain the radial rooted at λ_m
and the stellate structure of rays and island molecules beyond its tip.

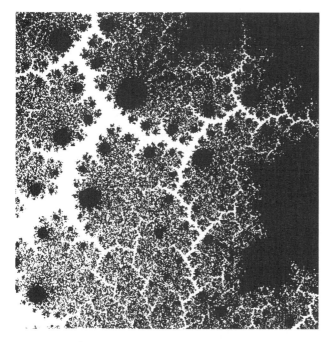

Figure 2 (parameter spaces). Detail of the M-set of $z \to \lambda z(1-z)$, when $-\log\lambda/2\pi i$ lies in a neighborhood of the golden ratio $(\sqrt{5}-1)/2$. The disc $|\lambda| < 1$ is visible to the top right.

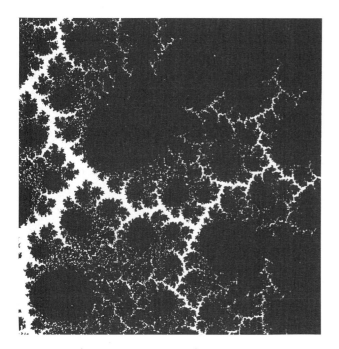

Figure 3 (parameter spaces). An expanded detail of Figure 2, in the region of its top right corner. The disc $|\lambda| < 1$ is again visible to the top right. This Figure is very analogous to Figure 2, but the white strips are narrower, which is a symptom of D = 2.

The striking overall resemblance between Figures 2 and 3 and Figure 1 provides the evidence for the above conjecture.

Discussion. What causes the behavior shown in Figure 2 and 3? To find out, let us compare the stellate structures of rays and island molecules that exist beyond the tip of each radical in the M-set. The orders of branching of these structures are the orders of bifurcation of the sequence that leads to the tip in question. When the radical is large, Paper III shows that the product of the orders of bifurcation is small, hence the largest among them is also small. Therefore, the large radicals' stellate structures have the appearance of sparse "trees". For example, the tip corresponding to an infinity of 2-bifurcations continues by a spear without brances. Similarly, FGN, p. 189 shows the tip corresponding to one 3-bifurcation and an infinity of 2-bifurcations: there is little branching. The same p. 189 of FGN shows, however, that one 100-bifurcation followed by an infinity of 2-bifurcation yields a much thicker structure. Increasingly small radicals have a rapidly increasing probability of otherwise very high bifurcations; as a result, their smaller structures become increasingly thick "bushes".

It is clear that the local shapes of the boundary of the M-set is dominated by an infinity of infinitely small and "bushy" structures. They crowd each other, hence the above conjecture.

Comment Fig. 1 was drawn 7 years before Figures 2-3. The latter exhibit small branches throughout, while in Figure 1 small branches only arise by division of slightly larger branches. Figure 1 is a variant of a fractal model of the lung; a more suitable background for the present analogy would have been provided by a fractal model of the vasculature!

Acknowledgments. Figure 1 was prepared by Sigmund W. Handelman. Figures 2 and 3, computed with programs written by V. Alan Norton, were prepared by James A. Given.

References: See the end of Paper VII.

17

On the Dynamics of Iterated Maps VI: Conjecture That Certain Julia Sets Include Smooth Components

Benoit B. Mandelbrot

IBM Thomas J. Watson Research Center
Yorktown Heights, New York

The Julia set $F*$ of an analytic map $z \to f(z)$ may be the boundary of an atom, of a molecule, or of a "devil's polymer" in the z-plane. Let $f(z) = z^2-\mu$, with $\mu \neq 0$, and denote the boundary of one of the atoms of $F*$ by H. When μ is the nucleus of a cardioid-shaped atom of the M-set, it is conjectured that the fractal dimension D_p of H is 1. Thus, H may be a rectifiable curve (of well defined length), or perhaps only a borderline fractal curve (of logarithmically diverging length).

1. THE MAP $z \to f*(z) = z^2-\mu$ AND ITS JULIA SETS $F*$.

The present paper develops an important remark made in passing, and not very explicitly, in Paper II (1983), p. 233. Then some generalizations are described.

For the quadratic the map $z \to z^2-\mu$, the Julia set $F*$ is the boundary of the set F of points that do not iterate to infinity under f*. When $\mu = 0$, this $F*$ is of course the unit circle and when $\mu = 2$, $F*$ is easily seen to be an interval. But for all other μ's, the $F*$ are unsmooth sets.

Of special interest are the μ's such that the set F has interior points. These μ's include those in the semi-open M-set (that is, the interior of the M-set, plus the bonds between atoms of the M-set.) The structure of F depends on µ and several possibilities will be explored.

When µ lies in the large atom of the continental molecule of the M-set, F is an atom; that is, $F*$ is a Jordan curve (simple loop).

When µ lies in the continental molecule of the M-set, but not in its large atom, F is a molecule. That is, F is the union of a (denumerably infinite) number of atoms, any two of which can be joined by a continuous curve that lies within F and crosses a finite number of other atoms.

When µ lies in an island molecule of the M-set, F has a "devil's polymer" structure, as described in Paper III when discussing the M-set.

A further possibility is examined in Paper VII.

In the second and third of the above possibilities, the complement of $F*$ includes a denumerable infinity of maximal open components that do not include the point at infinity. The boundary of one of these components is to be called H, and others are the preimages of H under f*(z). The set and its preimages being smooth images of each other, they have the same "partial" fractal dimension D_p, which at most equal to the dimension of $F*$. When a fractal set is the union of curves that are individually smoother than the whole, then (Mandelbrot 1982, p. 119) the overall fractal dimension is not a measure of the roughness of the whole, but of its fragmentation.

Fatou 1919 had shown that in general it is impossible for a subset of $F*$ to be an <u>isolated</u> analytic arc. Under wider conditions, $F*$ may include <u>non</u>-isolated analytic arcs. For example, when µ is real and the map f*(z) is chaotic, $F*$ includes a real interval and includes its preimages, which are smooth algebraic curves (see below). Now we proceed to observations that suggest a second such example.

2. OBSERVATIONS CONCERNING THE JULIA SETS   WHEN µ IS THE NUCLEUS OF A CARDIOD-SHAPED LARGE ATOM OF THE M-SET AND µ ≠ 0.

In this case, H can be defined as the boundary of the domain of immediate attraction of z=0. The origin may be called the nucleus of H, and the preimages of H have as nuclei the corresponding preimages of 0. When µ = 0, the set H is identical to $F*$ and is a circle. When µ ≠ 0, the observation is that H is nearly a disc, hence $F*$ is the union of near circles. They look so smooth as to seem to have tangents, and therefore to be rectifiable. It has been mentioned that they are linked in a "branched devil's polymer" structure. An example is shown on Figure 1.

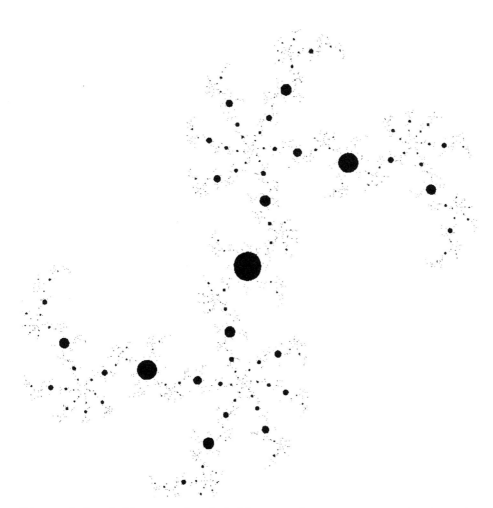

Figure 1 (variable space). A Julia set whose structure is that of a "devil´s polymer" made up of near circular mono-atomic molecules. The map is $z \rightarrow z^2 - \mu$, where μ is the nucleus of the largest atom of the largest island off a radical corresponding to 5-bifurcation. The cycle´s order is 6.

Special case: μ is real. In this case, the "branched polymer" structure of $F*$ is of special simplicity. Indeed, the real interval $[-z',z']$, with $z = 1/2 + 1/2(1+4\mu)^{1/2}$, is contained in F and can be said to form its "spine", in the sense that H and a collection of its preimages appear as "beads" strung along $[-z',z']$. Also, the preimages of $[-z',z']$ are smooth curves contained in F, and they form "ribs" along which other preimages of H are strung as "beads". At the origin 0 and at each preimage of 0, an infinity of ribs intersect, and near the origin these ribs are nearly straight. As a result, the preimages of H that are in the neighborhood of H lie along lines that radiate from 0.

Acknowledgements. Figure 1, computed with programs written by V. Alan Norton, was prepared by James A. Given.

References: See the end of Paper VII.

18

On the Dynamics of Iterated Maps VII: Domain-Filling ("Peano") Sequences of Fractal Julia Sets, and an Intuitive Rationale for the Siegel Discs

Benoit B. Mandelbrot

IBM Thomas J. Watson Research Center
Yorktown Heights, New York

Within the M-set of the map $z \to \lambda z(1-z)$, consider a sequence of points λ_m having a limit point λ. Denote the corresponding F^*-sets by $F^*(\lambda_m)$ and $F^*(\lambda)$. In general $\lim F^*(\lambda_m) = F^*(\lim \lambda_m)$, but there is a very important exception. In some cases, the sets $F^*(\lambda_m)$ do not converge to either a curve or a dust, but converge to a domain of the λ-plane, part of which is called the Siegel disc S and the rest is made of the preimages of S. In such cases, $F^*(\lim \lambda_m)$ is not the set $\lim F^*(\lambda_m)$, but its boundary. The intuitive meaning of this behavior is discussed and illustrated in terms of Peano curves, and a mathematical question is raised concerning the non rational and non Siegel λ's.

1. INTRODUCTION

Special cases of otherwise intuitive mathematical theories often seem to exhibit a "pathological behavior" that requires a technically difficult special treatment. For example, physicists find it hard to believe that there can be very concrete significance to functions F with the property that $\lim F(X_n)$ is completely different from $F(\lim X_n)$. The discs whose existence is shown in Siegel 1942 involve such a special case of the

243

theory of iteration of rational functions (and of other analytic functions). The argument and the illustrations in the present paper relate "Siegel discs" to Peano curves, hence ought to show that the Seigel disc behavior is in fact eminently reasonable. It could have been expected on physical grounds.

We consider the dynamics of certain extremely non-linear maps $z \to f(z)$ of the complex plane on itself, and separate the linear and nonlinear terms at the origin by writing the map in the form $z \to \lambda z + g(z)$, where $g(z)$ satisfies $g(0) = g'(0) = 0$. In order to strip down inessential complications, we stay with the quadratic case $g(z) = -\lambda z^2$.

2. BACKGROUND: PEANO CURVES.

The key to the argument resides in the concept of Peano "plane-filling curve", which the fractal geometry of nature (Mandelbrot 1982) retrieved from among the monster sets and showed to have great concrete importance. A "plane-filling curve" is really a sequence of ordinary curves C_n having an unexpected limit. In this instance, we need a slight variant: we assume that the C_n are loops with no double point (Jordan curves). The criterion of Peano behavior is that there exists a plane domain D such that every point P of D can be written as $P = \lim_{n \to \infty} P_n$, where P_n belongs to C_n. Thus, the limit of the C_n is not at all a curve, but is the whole domain D. However, in order to have a curve as limit, one may say that the limit of the C_n is the boundary of D, which is a curve to be denoted by C. In the classic examples of Peano, Hilbert and Sierpinski, D is a square. But D is a very irregular fractal curve in some more recent constructions; Mandelbrot 1982, p. 68-69 illustrates an example of mine, where C is the Koch snowflake curve.

The point of the present paper is that the same "Peano" behavior is characteristic of certain sequences of Julia sets. The fact that the C_n are Jordan curves is inessential; in fact, we shall also encounter Peano sequences of curves that have infinitely many double points.

3. SIEGEL DISCS

The map $z \to f(z)$ is said to have a Siegel disc, when there exists a bounded domain of the plane, open and connected, within which $z \to \lambda z + g(z)$ is essentially reduced to its linear term, that is, the map is equivalent to a rotation of angle θ. This means that a suitable deformation (a holomorphic function obtained as the solution of Schröder's equation) tranforms z into a variable ω in terms of which the map becomes $(\omega-\omega_0)\to(\omega-\omega_0)e^{2\pi i\theta}$. The circles centered on ω_0 transform into nested Jordan curves whose union covers the inside of the Siegel disc.

When a Siegel disc exists, λ is of the form $\lambda = \exp(2\pi i\theta)$ with θ a "Siegel number", meaning an irrational number satisfying certain conditions that are rather complicated but sufficiently mild for such numbers to be of unit measure on [0,1]. The rough idea is that a Siegel θ cannot be well approximated by a sequence of rational numbers. The angle of the rotation $z\to\lambda z$ being irrational, the iterates are ergodic.

How should one call the dynamics corresponding to f(z) when θ is a Siegel number? Depending on one's feelings, one may call it highly ordered or highly disordered.

In order to understand this behavior, I think the key step is to investigate a Siegel map $f = \lambda z-\lambda z^2$ as the limit of a suitable sequence of quadratic maps $_mf$. Iteration associates to each $_mf$ a shape called F^*-set or Julia set, which (Paper III) is the boundary of the domain that fails to be iterated to infinity. The Julia set of the limit map f is in general the limit of the Julia sets of the maps $_mf$. But when the limit f is a Siegel map, the Julia set of the limit map f is completely different from the limit of the Julia set of the approximating maps $_mf$. Indeed, the Julia set of the limit is a curve. But the limit of the Julia sets is not a curve, but a bounded domain D of the complex plane. This D includes a component called Siegel disc, and is the union of a Siegel disc and its preimages under f. The Julia set F^* of the limit f of the $_mf$ is the boundary of D.

4. APPROXIMATION   OF   THE   GOLDEN   λ = EXP(2πiσ) BY A SEQUENCE   OF "FIBONACCI BONDS" OR FINITE BIFURCATION VALUES OF λ.

Siegel 1942 implies that the Siegel members include the golden ratio σ, which is best defined as the continued fraction exclusively made of 1's, namely 1/(1+1/(1+1/(1+1/...))). Clearly, $\sigma^{-1} = 1+\sigma$, hence the equation $\sigma^2+\sigma=1$, whose root less than 1 is $1/2(-1+\sqrt5)$ = .6180339.

To gain an intuitive feeling of the action of the map $z\to e^{2\pi i\sigma}z(1-z)$, this action was obtained as the limit of the actions of two approximating sequences of maps $z\to\lambda_m z(1-z)$.

In a first study, λ_m was kept satisfying $|\lambda_m| = 1$, but $z\to e^{2\pi i\sigma}z(1-z)$ was approximated by $z\to e^{2\pi i\sigma}z(1-z)$, where σ_m is the rational number whose continued fraction expansion is made of m repetitions of 1. This is the

Figure 1

Figures 1 to 4 (variable space). The Julia sets of $z \rightarrow \lambda z(1-z)$ for four
values of the parameter λ. To demonstrate the nature of convergence to a
Siegel disc, the values of $\log\lambda_m/2\pi i$ are selected ratios of Fibonacci
numbers, that is, they are rational numbers that converge to the golden
ratio $(\sqrt{5}-1)/2$. The rational numbers used here are 2/3, 8/13, 34/55, and
144/233. The numbers of "petals" in these "flowers" are known to be the
denominators of the $\log\lambda_m/2\pi i$; unfortunately, these petals are not very
distinct on this Figure. See the Figures that follow for a method of
separating the petals.

mth Fibonacci number. By the general theory of the M-set M_λ of $z \rightarrow \lambda z(1-z)$,
each of these $\lambda_m = \exp(2\pi i\sigma_m)$ is a bifurcation point: the point of
contact between the atom $|\lambda| < 1$ and a smaller atom, the order of
bifurcation being $n(m)$ = denominator of σ_m.

Figures 2 to 4 show one half of the interior of the approximating
Julia sets for selected values of m. (The other half is obtained by
symmetry). A series of flower-like shapes seem to appear, their petal
numbers belonging to the Fibonacci series. At the same time, " wasp's
waists" appear.

5. <u>APPROXIMATION OF THE GOLDEN λ = EXP($2\pi i\sigma$) BY A SEQUENCE OF
   SUPERSTABLE λ_m*.</u>

The trouble with the preceding approximating sequence is that the petals
are not visibly separated from each other; to separate them, an
alternative approximating sequence was studied. It approximates $z \rightarrow e^{2\pi i\sigma}$

Figure 2

Figure 3

Figure 4

z(1-z) by $z \rightarrow \lambda_m^* z(1-z)$, where λ_m^* denotes the "nucleus" of the atom of period n(m) whose root lies at $\exp(2\pi i \sigma_m)$. This λ_m^* is a superstable value, meaning that the n(m)th iterate of $_m f$ satisfies $_m f_{n(m)}(0) = _m f_{n(m)}'(0) = 0$. Figures 5 to 7 show that the same petals that were squeezed together in the case of the sequence λ_m have become well-separated.

6. A MATHEMATICAL QUESTION.

The irrational numbers θ split into two classes: the Siegel numbers, which are the great majority, and all the other numbers. Supposing that θ_m converge to a non-Siegel number, let us attempt to visualize the behavior of the Julia sets F^*_m. This attempt is simplified by concentrating on the set Q_m defined as the intersection of F^*_m with a suitable circle. In the Siegel case, the fact that F^*_m converges to a domain filling curve, implies that the set Q_m converges to the whole circle. When θ converges to a rational number, F^*_m is not a domain-filling sequence, and Q_m converges to a Cantor dust of zero measure.

The preceding observations lead to the following query: when θ is neither rational nor a Siegel number, what is the behavior of Q_m? Does it perhaps converge to a Cantor dust of dimension 1, or even (for some $\theta's$)

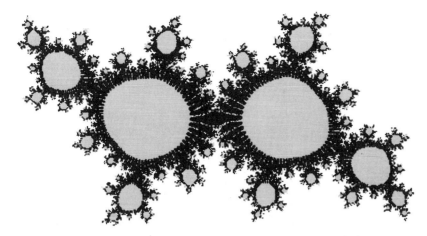

Figure 5 (variable space). In order to separate the petals, a λ_m in the Fibonacci sequence, namely $\exp[2\pi i(55/89)]$, is replaced by the nucleus of the radical rooted at λ_m. This Figure gives an overall view of the corresponding Julia set. The white, black and gray zones correspond, respectively to values of z_0 that converge to infinity, converge to a cycle, or have failed to converge when the calculation was stopped. Each petal extends to an unstable fixed point in the middle of the gray zone.

of positive linear measure? If the latter is the case, F^* would be of positive planar measure!

7. SIEGEL DISC AS LIMIT OF FRACTAL DUSTS.

In section 4(resp. in section 5), a λ-value on the boundary of the M-set of f(z) is approximated by λ_m's that lie on the boundary of the M-set (resp., inside it). Now let the approximate values be chosen outside of the M-set. The corresponding Julia sets are dusts. When the approximants converge to a rational $\lambda = \exp(2\pi i m/n)$, these dusts coalesce into curves. However, convergence of λ_m to a Siegel λ results in a dust coalescing into a domain.

8. HISTORY

"Siegel" discs were originally described in Julia 1918; see Julia 1968 I pp. 311-317. Next, Julia claimed to prove they could **not** occur (Julia 1968 I p. 321); he later recognized his proof was flawed, and finally dropped the topic (Julia 1968 I, starting p. 21 bottom). Siegel 1942

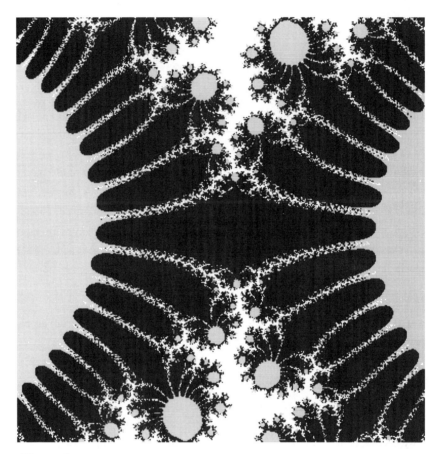

Figure 6

Figures 6 and 7. These Figures are detailed views of the neighborhood of
the origin in the preceding Figure 5, and of the analogous figure for the
next Fibonacci ratio, $\log\lambda/2\pi i = 89/144$. As the golden ratio is
approached, the petals become increasingly thin, and their outline
converges to the interior of a domain of the plane.

Figure 7

proved their existence for a certain set of irrational values of θ, having a measure equal to 1. This set has since been broadened.

ACKNOWLEDGEMENTS. I am in debt to Gregory and David Chudnovsky for interesting comments concerning Siegel discs, and to V. Alan Norton and James A. Given for many interesting discussions as well as for the computer programs used to draw the illustrations in this paper.

References

Note that this is strictly a list of references, and not a bibliography of a field of mathematics that has become very active again since 1980, and now experiences explosive growth.

Collet, P. and Eckman, J. P. 1980. Iterated Maps on the Interval as Dynamical Systems. Boston: Birkhauser.

Cvitanovic, P. and Myrheim, J. 1982. Universal for period n-triplings in complex mappings, Physics Letters **94**A, 329-333.

Douady, A. 1982. Systems dynamiques Holomorphes. Séminaire Bourbaki n°599 (Available from Société Mathématique du France, as an issue of Astérisque).

Douady, A. And Hubbard, J. 1982. Iteration des polynomes quadratiques complexes. Comptes rendus (Pairs) **294**-I, 123-126.

Hardy, G. H. and Wright, E. M. 1960. An Introduction to the Theory of Numbers. (4th edition) Oxford: Clarendon Press.

Fatou, P. 1906. Sur les solutions uniformes de certains équations fonctionelles. Comptes rendus (Paris) **143**, 546-548.

Fatou, P. 1919-1920. Sur les équations fonctionnelles. Bull. Société Mathématique de France **47**, 161-271; **48**, 208-314.

Julia, G. 1918. Mémoire sur l'itération des fonctions rationnelles. J. de Mathématiques Pures et Appliquées 4: 47-245. Reprinted (with related texts) in Oeuvres de Gaston Julia, Paris, Gauthier-Villars. 1968, 121-319.

Mandelbrot, B. B. 1980. Fractal aspects of the iteration of $z \rightarrow \lambda z(1-z)$ for complex λ and z. Non Linear Dynamics, Ed. R.H.G. Helleman. Annals of the New York Academy of Sciences, 357, 249-259.

Mandelbrot, B. B. 1982. The Fractal Geometry of Nature, New York: W. H. Freeman.

Mandelbrot, B. B. 1983. On the quadratic mapping $z \rightarrow z^2 - \mu$ for complex μ and z: the fractal structure of its -set and scaling. Physica **7D**, 224-239; also in Order in Chaos. Ed. D. Campbell and H. Rose, Amsterdam, North Holland.

Mandelbrot, B.B. 1984. On the dynamics of iterated maps VIII: The map $z \rightarrow \lambda(z+1/z)$ from linear to planar chaos, and the measurement of chaos. Chaos and statistical Mechanics, Ed. Y. Kuramoto, New York: Springer.

Ruelle, D. 1982. Repellers for real analytic maps. Ergodic Theory and Dynamic Systems, 2, 99-107.

Siegel, C.L. 1942. Iteration of analytic functions. Annals of Mathematics 43, 607-612.

Additional References in the Second Printing.

Blanchard, P. 1984. Complex analytic dynamics of the Riemann Sphere. Bulletin of the American Mathematical Society (N.S.) 11, 85-141.

Mandelbrot, B.B. 1985. On the Dynamics of Iterated Maps IX: Continuous Interpolation of the complex discrete map $z \rightarrow \lambda z(1-z)$ and related topics. Physica Scripta, 79, 59-63.

Peitgen, H. O. and Richter P. 1985. Frontiers of Chaos (Catalog of an Art Exhibit) To be published as Beauty of Fractals by Springer-Verlag in 1986.

Author Index

Subject Index